Prepare Yourself
For the
Professional Practice Examination

A. R. MEMARI, Ph.D., P. Eng.

TABLE OF CONTENTS

PREFACE

Practice of engineering as professional was regulated for the first time in 1887, when the Canadian Society of Civil Engineers held its first meeting, which by 1918, it changed its name to Engineering Institute of Canada with the goal to improve the professional engineering practice. Engineering Institute of Canada drafted an act, which was approved by 77 percent of its members. In 1920, licensing law was passed. According to this law all the engineers were required to have license to practice engineering.

From the very beginning, the basic aim of CSCE was to establish high standards for engineers and to improve the practice of professional engineering. It required all of the applicants to be at least thirty years of age and to be in possession of at least ten years of experience.

It is mandatory for all the applicants regardless of their academic backgrounds to write the Professional Practice Examination, which is a three-hour closed book examination in Engineering Law and Ethics.

This is a first book of its kind that provides the PPE candidates with answers to the questions of the previous PPE. Attempts have been made to produce this book in a very simple and easy to understand book, so the complicated language used in the two textbooks has been mitigated. In the first six chapters question papers of Part A and Part B of years 2003 and 2004 have been answered in the simplest and informal language. In order to test your readiness for PPE, chapter seven comprises only questions of the two Parts. This book is an excellent prerequisite to the two textbooks, which provides the PPE candidates with all what they should know before attending the examination. Attend the PPE with integrity and confidence

A.R. Memari, Ph.D., P. Eng.

ABOUT THE AUTHOR

The author of this E - book is a holder of Ph.D. in High Voltage Engineering and has been engaged in Post Doctoral Research Work at University of Toronto. Dr. Memari has published scientific papers in the area of magnetic field associated with high voltage transmission line, which have been utilized by inventors, scientists and even by the Ph.D. candidates for their Ph.D. Theses all over the world.

Dr. Memari is also author of an E-book, " Advanced Approach to Mitigate Magnetic Fields and Your Health.", which is published by the Bentham Science Publisher.

Dr. Memari is a Professional Engineer Ontario and is in possession of a patent with Canadian Intellectual Property Office, no. 2585773 and has been acting as a paper reviewer for IEEE Transactions and foreign scientific journals.

CHAPTER 1

PPE, Part "A", Part "B", April 2003

Abstract. This chapter discusses in a very informal manner the suggestive answers of the author to the question papers of April 2003. The two parts (Part "A" – Professional Practice and Ethics) and (Part "B" – Engineering Law and Professional Liability) of Professional Practice Examinations have been completely answered. The simplicity of the language adopted in this chapter will act as a useful tool for the PPE candidates to further understand these two subjects.

ASSOCIATION OF PROFESSIONAL ENGINEERS OF ONTARIO

PROFESSIONAL PRACTICE EXAMINARTION-April 26, 2003

PART "A" – Professional Practice and Ethics

This examination comes in two parts (Part "A" and Part "B") which must be completed within the allotted **three-hour** time period. Both parts of the examination must be passes at the same sitting in order to be assigned a "PASS" for the Professional Practice Examination. The result of the combined examination, Part "A" plus Part "B", will be declared a "PASS" or "FAIL"

Use the correct colour – coded Answer Book for each part, place in the correct envelope and **seal after completed.**

> **White Answer Book for Part A white question paper:**
> **Coloured Answer Book for Part B coloured question paper:**

This is a "**CLOSED BOOK**" examination. No aids are permitted other than the excerpts from the 1990 Ontario Regulation 941 covering sections 72 (Professional Misconduct) and 77 (Code of Ethics) supplied at the examination. Dictionaries are **not** permitted.

The marking of the questions will be based not only on academic content, but also on legibility and the ability to express yourself clearly and correctly in the English language. If you have any doubt about the meaning of a question, please state clearly how you have interpreted the question.

All **four** questions constitute a complete paper for Part "A", Question 1 is worth 40 marks. Question 2, 3 and 4 are each worth 20 marks.

Where a question asks if a certain action by an engineer was ethical or not, a simple "yes" or "no" answer is not sufficient. You should identify the applicable clauses in Regulation 941 and comment on the action in each situation.

Page 1 of 4

1.1

Part A – PPE, April 26, 2003   page 2 of 4

Question # 1

(i)

Professional Engineers Ontario's (PEO) *Code of Ethics* sets out the expected professional behaviour and conduct for PEO members and its other licensees.

(15 marks)   (a) Discuss the compatibility of PEO's *Code of Ethics* with the goals of industry and the role of the professional engineer in dealing with any incompatibility between the two, if any.

(5 marks)   (b) What is the relationship between PEO's *Code of Ethics* and the Definition of Professional Misconduct?

(ii)

(5 marks)   (a) What is PEO's principal object?

(15 marks)   (b) List and describe the three main functions that PEO performs towards the fulfillment of its legislated mandate.

Question # 2

(20 marks)   You are a licensed (mechanical) professional engineer charged with enhancing the efficiency of a liquid detergent production line for your employer, a soup manufacturer. During your work you have access to confidential company information and observe that the company is adding very small quantities of a well-known carcinogen (i.e., a substance suspected of causing cancer) to the detergent but it is not listing it as an ingredient. This confidential information is irrelevant o your work. However, you are aware that the additive is a banned substance.

Under the circumstances, what action(s) are you obliged to take as a professional engineer?

Question # 3

(20 marks)

Professional engineer Epsilon was sent by the Canadian firm, ABC international Inc., to serve as resident geological engineer in Puere, South America. ABC International Inc. had been hired by the Puerean government to oversee a project being undertaken by another Canadian company, XYZ Overseas Inc. The project involved the construction of a 400 Km highway across a mountainous region.. Although relatively new to ABC International Inc. Epsilon, with more than twenty-five years of experience, was given the key assignment of ensuring that contract agreements between the Puerean government and XYZ Overseas Inc. were met.

Epsilon's signature on the payroll certifies that the interests of the Puerean government were being served. Almost immediately Epsilon began to experience doubts about the project. The design for the highway, which as it turned out, was originally done by ABC International Inc. called for cutting deep channels – some of them more than 100 meters through the mountains with cliff rising sharply on both sides of the road. Epsilon was concerned that with the instability of the mountains, it did not appear as if enough geological borings had been taken to identify potential slide areas. Epsilon fear were confirmed, unfortunately, when several slides and other construction accidents occurred killing some workers. XYZ Overseas Inc. asked Epsilon to add to the payroll to cover the substantial costs for slide removals.

Epsilon viewed the request as one of "padding" and, therefore, not justified by anything in the contract. At first, Epsilon's position was supported by Epsilon's firm, however, with mounting pressure by XYZ Overseas Inc., ABC International Inc. ordered Epsilon to add the slide removal costs to the payroll. Epsilon refused to do so, insisting that it would be a violation of the Puerean government's interests which ABC International Inc. was charged to protect. Epsilon was relieved of the resident engineer's responsibility and was subsequently dismissed by ABC International Inc.

Discuss Epsilon's actions, as well as those of ABC International Inc. and XYZ Overseas Inc. in terms of Professional Engineers Ontario's Code of Ethics. Is there a recommended recourse that Epsilon might pursue in view of the dismissal?

Question # 4

Tau, P. Eng. employed by a large well-known automobile parts testing laboratory, represents the firm on an international standards advisory committee. Eight members of the ten-person committee are licensed professional engineers. After an extensive discussion on a standard at a recent meeting, the committee voted in favour. Although Tau was the only dissenting vote, the committee considered Tau's reasons for objecting; and after further discussions, it agreed that they were unsubstantiated and passed a motion to accept and publish the standard.

(20 marks)

Subsequently, the laboratory where Tau is employed received a contract to test automobile parts to this standard and Tau was assigned the job of supervising the tests, preparing the final report indicating that the parts meet the standard, and signing it on behalf of the firm. However, Tau is still vehemently opposed to the standard and therefore refuses the assignment. Tau argues that signing a report attesting to the conformance of the parts to the standard would suggest endorsement of it.

Is Tau's argument for refusing to undertake the assignment correct? What action might Tau have taken?

1.2 Suggestive answers to Ethics, APRIL 2003

Answer to question 1

(i)-(a)   In accordance with the Code of Ethics, all the professional engineers are duty bound to act at all times with loyalty and honesty towards their employers, public, clients and friends. A professional engineer must consider the welfare of the general public as paramount.

A professional engineer must always act faithfully towards his / her employer and the professional engineer is obligated to keep the information, technical method and business affair of his / her work place confidential and must avoid any conflict of interest that might influence his / her judgment.

If due to any reason, a professional engineer is moonlighting he / she must disclose it to his / her employer and he / she must be sure that it does not conflict with his / her main job. The professional engineer must always cooperate with his / her colleagues. He / she must not review work of other colleagues without their consents unless the practitioner is no longer working for the same company. Under no circumstances, a professional engineer is permitted to accept bribe of commission. He must maintain the integrity of his profession and at all times he must remain impartial when his professional opinion is required.

Even though a professional engineer is duty bound to comply with his /her employer's instructions but the professional engineer must regard the well fair and safety of the general public as paramount, therefore, whenever the safety or interest of the public is in danger, the professional engineer must seriously protest. He may even blow the whistle.

(i)- (b)   Under the Professional Engineering Act of Ontario, the Code of Ethics provide for high standard of duty, conduct and integrity. The main purpose of the Association is to protect welfare of the general public. Professional misconduct is defined as negligence and incompetence. According to this Act, the Association is empowered to suspend, or expel members who are guilty of professional misconduct.

(ii)-(a)   The principal objective of PEO is to regulate the practice of professional engineering , to govern its members, to serve and protect the welfare of the general public.

(ii)-(b)   In order to administer its provincial Act, province of Ontario has established its self governing Association of Professional Engineers Ontario. This Association has developed regulations, bylaws and Code of Ethics.
Regulations are set of rules that implement or support the Act. Regulations deal with qualification for admission to the Association, professional conduct and disciplinary procedure.
Bylaws are rules that administer the Association. Bylaws deal with election procedure financial matter and meetings.
The Code of Ethics is rules of personal conduct. Regulations, bylaws and the Code of Ethics have been made under the authority of each Act.

Answer to question 2

You, as an employee of this firm, have a duty to keep your employer's business confidential. Pursuant to the Code of Ethics, you, as a professional engineer, are duty bound to safeguard welfare of the public and consider safety of general public as paramount. In the given case, a hazardous substance endangering health of the public has been used. You have an ethical duty to take an immediate action. Your first step should be to inform the concern authority of your firm about the substance in writing. If the firm were honest, they would take immediate action to remove the substance. If the firm were dishonest, then the matter would be serious. If due to consumption of the detergent a serious event occurs, then the dangerous substance would enter public domain. You would be probably investigated for possible unethical behavior, or even collusion with the firm. A professional engineer who discovers that his / her employer is not honest must immediately dissociate himself / herself from any action contrary to the Code of Ethics. You may even blow the whistle if the firm's authority completely ignores your opinions as well as your instructions.

Answer to question 3

Courts always enforces those terms and conditions that are clearly written in the contract. Epsilon realizes that the request for adding to payroll to cover the cost of slide removal is not consistence with the terms of the contract. Consequently, he has done a very good job to refuse to accept the cost of slide removal. Epsilon is a professional engineer and in accordance with the Code of Ethics he is duty bound to act with integrity and honesty. Epsilon is acting as an agent for the Puerean government, therefore a relationship of trust must exist between these two. Epsilon was ultimately dismissed by ABC International Inc, due to the mounting pressure by XYZ Overseas Inc. Such a dismissal is defined as wrongful dismissal and has a potential to end up in the court.

Answer to question 4

A professional engineer, in some occasion, may be instructed by his / her employer to involve himself / herself in performing something that the engineer may find it morally wrong. An engineer may be asked by his / her employer to do something that is neither illegal nor contrary to the Code of Ethics, but it, simply, contradicts the engineer's conscience. In this situation, the engineer must define the ethical problem as clearly as possible. Obviously, ethical problem may be associated with special consequences. The engineer must view the problem from the employer's point of view. Refusing to comply with your employer's instructions may result in disciplinary action or dismissal. In the given case, professional engineer, Tau, was assigned to a type of job that contradicted his / her moral. Even though, Tau can refuse to accept the assignment, but he / she must also be alert of possibility of being dismissed.

ASSOCIATION OF PROFESSIONAL ENGINEERS OF ONTARIO

PROFESSIONAL PRACTICE EXAMINARTION-April 26, 2003

PART "B" – Engineering Law and Professional Liability

This examination comes in two parts (**Part "A" and Part "B"**) which must be completed within the allotted **three-hour** time period. Both parts of the examination must be passes at the same sitting in order to be assigned a "PASS" for the Professional Practice Examination. The result of the combined examination, Part "A" plus Part "B", will be declared a "PASS" or "FAIL"

Use the correct colour – coded Answer Book for each part, place in the correct envelope and **seal after completed.**

> *White Answer Book for Part A white question paper:*
> *Coloured Answer Book for Part B coloured question paper:*

This is a "**CLOSED BOOK**" examination. **No** aids are permitted other than the excerpts from the 1990 Ontario Regulation 941 covering sections 72 (Professional Misconduct) and 77 (Code of Ethics) supplied at the examination. Dictionaries are **not** permitted.

The marking of the questions will be based not only on academic content, but also on legibility and the ability to express yourself clearly and correctly in the English language. If you have any doubt about the meaning of a question, please state clearly how you have interpreted the question.

All **four** questions constitute a complete paper for Part "B", Question 1 is in two parts totaling 40 marks, including a choice of four definition type questions of equal value[1(a)] and four separate questions of equal value **[1(b)].** . Questions 2, 3 and 4 are each worth 20 marks.

1.3
(MARKS)

(40)   1.   Please not that this question 1 is a two-part question, worth 40 marks in total.   Question 1(a) is a definition type question, worth 20 marks. Question 1(b) consists of four parts of equal value, totaling 20 marks.

(20)   (a)   <u>Briefly</u> define and explain any <u>four</u> of the following:

   (i)   Mitigation of damages
   (ii)   Gratuitous promise
   (iii)   Parol evidence rule
   (iv)   Penalty clause
   (v)   Consequential damages
   (vi)   Secret Commission

(20)   (b)   (i)   can there be a difference between an engineer's mistake or error and an engineer's negligence? Is it possible that an engineer might make a mistake that results in damages but not be liable for negligence? Discuss and explain.

   (ii)   What is meat by a "limitation period"? In your explanation, please provide three examples of limitation periods relevant to engineers and contractors.

   (iii)   Explain the basis of the formation of "Contract A" in the tendering process. Discuss what an owner can do to avoid Contract A problems arising

   (iv)   In some construction contracts, an engineer is authorized to be the sole judge of the performance of work by the contractor. Where such a provision is stated, is it possible that the provision will not be enforceable on account of the manner in which the engineer performs his or her duties? Explain.

(20)   2   A telecommunication development company leased an outdated and unused underground pipe system from an Ontario municipality. The developer's purpose in leasing the pipe was to utilize it as an existing conduit system in which to install a fibre optic cable system to be designed, constructed and operated in the

municipality by the telecommunication developer during the term of the lease. All necessary approvals from regulatory authorities were obtained with respect to the proposed telecommunications network.

The telecommunication development company then entered into an installation contract with a contractor. For the contract price of $ 4,000,000, the contractor undertook to complete the installation of the cable by a specified completion date. The contract specified that time was of the essence and that the contract was to be completed by the specified completion date, failing which the contractor would be responsible for liquidated damage in the amount of $50,000 per day for each day that elapsed between the specified completion date and the subsequent actual completion date. The contract also contained a provision limiting the contractor's maximum liability for liquidated damages and for any other claim for damages under the contract to the maximum amount of $ 1,000,000.

Due to its failure to properly staff and organize workforce, the contractor failed to meet the specified completion date. In addition, during the installation, the contractor's inexperienced workers damaged significant amount of the fibre optic cable, with the result that the telecommunications development company, on subsequently discovering the damage, incurred substantial additional expense in engaging another contractor to replace the damaged cable. Ultimately, the cost of supplying and installing the replacement cable plus the amount of liquidated damages for which the original contractor was responsible because of its failure to meet the specified completion date, totaled $ 1,800,000.

Explain and discuss what claim the telecommunication development company could make against the contractor in the circumstances. In answering, explain the approach taken by Canadian courts with respect to contracts that limit liability and include a brief summary of the development of relevant case precedents.

(20)   3.   An information technology firm submitted a bid to design and install software and hardware for an electronic technology process to control the operation of large scale sorting equipment for a major international courier company.

The firm's fixed guaranteed maximum price was the lowest bid and the contract was awarded to it. The contract conditions entitled the information technology firm to terminate the contract if the courier company did not pay monthly progress payments within 15 days following certification that a progress payment was due. Pursuant to the contract, the certification was carried out by an independent engineering firm engaged as contract administrator.

The work under the contract was to be performed over a 5 months period. After commencing work on the project the information technology firm determined that it had made significant judgment errors in arriving at its bid price and that it would face a major loss on the project. Its concern about the anticipated loss was increased further when it also learned that, in comparison with the other bidders, its bid price was extremely low and that, in winning the bid, it had left more than one million dollars "on the table".

Two monthly progress payments were certified as due by the independent engineering firm and paid by the courier company in accordance with the terms of the contract. However, after the third monthly progress payment was certified as due by the independent engineering firm, the courier company's finance department asked the information technology firm's representative on the project for additional information relating to an invoice from a subcontractor to the information technology firm. The subcontractor's invoice comprised a portion of the third progress payment amount. The courier company's finance department requested that the additional information be provided prior to payment of the third progress payment.

There was nothing in the signed contract between the information technology firm and the courier company that obligated the information technology firm to provide the additional information on the invoice from its subcontractor. However, the information technology firm's representative did verbally indicate to the courier company's finance department that the additional information would be provided.

The additional information relating to the subcontractor's invoice was never provided by the information technology firm.

Sixteen days after the third progress payment had been certified for payment, the information technology firm notified the courier company in writing that it was terminating the contract because the courier company was in default of its obligations to make payments within fifteen days pursuant to the express wording of the contract.

Was the information technology firm entitled to terminate the contract in these circumstances? In giving reasons for your answer, identify and explain the relevant legal principle and how it would apply.

(20)   4.   Live Rail Inc. ("Live Rail), a company specializing in the manufacture and installation of railway commuter systems was awarded a contract by a municipal government to design and build a transit facility in British Columbia. The contract specified electrically powered locomotives. As part of the design, Live Rail was contractually obligated to design an overhead contact system in a tunnel. Live Rail subcontracted the sub design of the overhead contact system to a consulting design firm, Ever Works Limited ("Ever Works").

Ever Works designed an overhead contact system in the tunnel, however, in doing so it did not carry out any testing nor did it gather any data of its own relating to the conditions inside the tunnel. It did not even request copies of underlying reports which, had been examined, would have indicated that there was a large volume of water percolating through the tunnel rock and that the tunnel rock contained substantial amounts of sulphur compounds. The project documentation that was turned over to Ever Works by Live Rail did not include underlying reports, but did identify the existence and availability of the underlying reports.

The construction of the rail system through the tunnel was completed in accordance with Ever Work's design. However, within eight months of completion, the overhead contact system in the tunnel became severely corroded and damaged due to the water seepage in the tunnel.

As a result of the corrosion damage, the municipality had to spend substantial additional money on redesigning and rewiring the system.

What potential liabilities in tort law arise in this case? In your answer, explain what principles of tort law are relevant and how each applies to the case. Indicate a likely outcome to the matter.

1.4   Suggestive answers to Law, APRIL 2003

Answer to question 1

(a)

(i)   When a party to a contract sustains damages through breach of contract, he must take reasonable steps to mitigate the damages, otherwise his negligent conduct will be taken into account when fixing award for the damages.

(ii)   It is a promise without consideration. A promise or a contract without consideration cannot be enforced. An exception exists in the case of equitable estoppel .

(iii)   "Parol" means verbal. It is a rule, which states that all the terms and conditions, which are mutually agreed upon must be mentioned in the contract, otherwise they cannot be enforced. This rule precludes evidence of omitted terms.

(iv)   A contract often contains a provision requiring a party to the contract to pay for prescribed damages, if a certain event occurs. These pre-estimated damages must be genuinely drafted, otherwise the court may not enforce the term.

(v)   Consequential damages are also known as indirect damages. When a contractor is working on an owner property and due to his negligent services the adjacent land or firm sustains damages, then indirect or consequential damages have occurred. There is a possibility that an owner may be fined due to the contractor's non-compliance with the environmental protection.

(vi)   It is money, or something of value that has been offered through a secret agreement to show a favour or disfavour to a third party. Secret commission is an offence under Criminal Code of Canada. If a Professional Engineer is involved in secret commission, then he/she will be charged under Regulation 941/90.

(b)

(i)   Negligence is defined as an act or omission in carrying out of the work of a practitioner that constitutes a failure to maintain the standard that a reasonable and prudent practitioner would maintain in the circumstances. Negligence is a result of careless performance of an engineer. Error is one that does not constitute negligence. An engineer does what ever he/she can do to have the job done. Since Canadian courts do not disregard exemption clauses that are clearly written, then engineers have the opportunity to add an exemption clause disclaiming any responsibility to the third party. The engineer can also limit his / her liability.

(ii)   Limitation period is a specified period of time during which a claim for the damages can be made. If a claim is made after expiry of the period, then the claim would not succeed. This is known as "statute barred".

(iii)   There are two types of construction contracts. Contract A and contract B. Contract

A is an irrevocable contract so it must have a consideration. When a bidder submits his bid in response to a call for bid, contract A is formed. In order for an owner to avoid problems associated with contract A, he must clearly inform bidders that in dealing with them, he has preserved the right of as much flexibility as possible

(iv)   When an engineer is acting as a sole judge, he must perform his duty with care, skill and absolute honesty, otherwise the penalties may be severe. If his dishonesty involves fraud, then the contract can be repudiated and the deceived party may be awarded for the tort of deceit.

Answer to question 2

This case is dealing with fundamental breach of contract. The contract between the telecommunications development company and contractor indicates that there would be a daily-liquidated damage of $ 50,000. The contract also includes an exemption clause limiting total liability of contractor to $ 1,000,000.

At one time Canadian courts followed English courts precedent in applying what was called the "fundamental breach doctrine". According to this doctrine, an exemption clause cannot be enforced in case of a fundamental breach. Eventually, this doctrine was overruled by English courts. Even though, the concept of this doctrine is not completely overruled by Canadian courts, but strong preferences are given to the exemption clauses in commercial cases. Canadian courts following English courts do not disregard the exemption clauses. In fact parties to the contract are free to select and choose their own terms and conditions. Provided they follow the five essential terms making a contract enforceable.

In the given case, the cost of supplying and installing plus the amount of liquidated damages because of the contractor's failure to meet the specified completion date is equal to $ 1,800,000. The contractor has also limited his liability to $1,000,000. Since Canadian courts do not disregard the exemption clauses that are genuinely written, the telecommunication company would be entitled to receive only 1,000,000, as mentioned in the exemption clause.

Answer to question 3

This is the case, which involves "equitable estoppel", which states that when a party to a contract makes a gratuitous promise, a promise without consideration, and the second party to the contract reasonably relies on it, then it would be unfair and inequitable to allow the first party to revert to his contractual terms. Court always enforces only those terms and conditions that are clearly and unambiguously written in a contract and a contract without consideration cannot be enforced. If a term has verbally been agreed upon, but has not been entered into the contract, then this term is not enforceable. But in case of a gratuitous promise, since it would be unfair and inequitable to let the first party return to his strict contractual term, the concept of equitable estoppel is well applicable.

In the given case, a representative of the information technology firm made a gratuitous promise, when the representative verbally indicated to the courier company's finance department that additional information would be provided. In this case the concept of equitable estopple is well applicable and information technology firm must be estopped. It would be unfair and inequitable to allow information technology firm to return to his strict contractual term.

Answer to question 4

This case is known as "concurrent liability in tort and contract". As much as a contractor is required to comply with the contract terms and conditions, he must also perform his services skillfully and with diligent and care. Simultaneous lack of these two, entitles the contractor to be concurrently liable to tort and contract. The non-defaulting party is entitled to some compensation. In the given case, Ever Work was assigned to design an overhead contact, but the case indicates that the municipality had to spend additional money on redesigning and rewiring the system. Consequently the contract has become purposeless. The municipality had to redesign and rewire the system, which is an indication that the system built by Ever Works did not serve the purpose for which it was supposed to be built. This is a non-compliance with the contractual terms.

The main purpose of tort is to compensate a party who has sustained damages as a result of negligent conduct of the second party. Claim for tort may arise, even though the two parties are not in contractual relationship. In order to be successful in tort claim, the plaintiff must prove that:
  1- the defendant owed the plaintiff a duty of care
  2- the defendant breached that duty by his negligent conduct
  3- the defendant conduct caused damages to the plaintiff
Ever Works had a duty to render his services with skill, care and diligence. Ever Works knew or ought to have known that municipality would rely on his performance. Ever Works ought to have foreseen the damages due to his negligent conduct. He did not carry out any testing nor did it gather any data related to the condition inside the tunnel. There was a report, though not been given to him, indicating that there was a large volume of water. Water seepage in the tunnel was the main reason for the damage to occur. This could have been avoided, had Ever Works requested for the copy and had conducted tests. Consequently he has breached his duty and municipality sustained damages. The prerequisite for tort claim can arise when damage manifests itself. In this case the damages are, corrosion, rewiring and redesigning the system. The contract does not indicate if there were an exemption clause to limit Ever Work's liability. Therefore relying on the information provided here, it is likely that municipality would be entitled to recover the reasonable costs of rewiring and redesigning the system as well as cost of future replacement of the wire.

CHAPTER 2

PPE, Part "A", Part "B", August 2003

Abstract. This chapter deals with the answers to the questions of question papers Part "A" Professional Practice and Ethics, and Part "B" – Engineering Law and Professional Liability of August 2003. In order to establish easy way to understand these two subjects, all the questions are answered in very informal way.

ASSOCIATION OF PROFESSIONAL ENGINEERS OF ONTARIO

PROFESSIONAL PRACTICE EXAMINARTION-August 9, 2003
PART "A" – Professional Practice and Ethics

This examination comes in two parts (**Part "A" and Part "B"**) which must be completed within the allotted **three-hour** time period. Both parts of the examination must be passes at the same sitting in order to be assigned a "PASS" for the Professional Practice Examination. The result of the combined examination, Part "A" plus Part "B", will be declared a "PASS" or "FAIL"

Use the correct colour – coded Answer Book for each part, place in the correct envelope and **seal after completed.**

> *White Answer Book for Part A white question paper:*
> *Coloured Answer Book for Part B coloured question paper:*

This is a "**CLOSED BOOK**" examination. **No** aids are permitted other than the excerpts from the 1990 Ontario Regulation 941 covering sections 72 (Professional Misconduct) and 77 (Code of Ethics) supplied at the examination. Dictionaries are **not** permitted.

The marking of the questions will be based not only on academic content, but also on legibility and the ability to express yourself clearly and correctly in the English language. If you have any doubt about the meaning of a question, please state clearly how you have interpreted the question.

All **four** questions constitute a complete paper for Part "A", Question 1 is worth 40 marks. Question 2, 3 and 4 are each worth 20 marks.

Where a question asks if a certain action by an engineer was ethical or not, a simple "yes" or "no" answer is not sufficient. You should identify the applicable clauses in Regulation 941 and comment on the action in each situation.

Page 1 of 4

2.1

Part A – PPE, August 9, 2003   **Page 2 of 4**

Question # 1 (40 Marks in Total)

(i) 5 marks for each of (a), (b), (c) and (d)

(a)   The practice of sealing (or stamping) an engineer document exposes a
P. Eng to liability. Why is it not a good practice to release drawings
which bear a photocopy of the professional engineer's stamp and signature?

(b)   For what purpose is PEO's *certificate of authorization issued?* What are
PEO's eligibility requirements for obtaining a *certificate of authorization?*

©   Are PEO's limited licence holders entitled to use the title "Professional
Engineer" or "P. Eng."? Explain.

(d)   There are four requirements that an applicant must meet in order to qualify
to be a consulting engineer. List two of them,

(ii) 10 marks for each of (a) and (b)

(a)   Explain the concept of *self - regulation* as it pertains to the engineering
profession in Ontario.

(b)   Describe the role of the professional engineer in society and the obligations
that accompany licensure.

Part A – PPE, August 9, 2003   **Page 3 of 4**

Question # 2 (20 Marks)

You are a licensed professional engineer and Chief Director of a federal government contract – granting agency. Prior to taking on your new position, you were a partner in a very successful consulting engineering firm. At the end of the summer last year, you sold your interest in the partnership to your partner, who is also a professional engineer, and soon thereafter you accepted the appointment to your current position. A few days later, much to your surprise, you learned that your former partner sold the firm to Octopus Enterprises, Inc., and became an officer of the cooperation. It is now several months later. In your new capacity, you are presented with documents recommending that you approve the award of a major engineering contract to Octopus.

In light of PEO's Code of Ethics, discuss your ethical obligations, if any, pertaining to the situation at hand.

Question # 3 (20 Marks)

Alpha, a consulting professional engineer, is retained by a client to design and supervise the construction of a warehouse. Following the completion of this work, Alpha is asked to provide similar professional services by another client who requests that alpha's fee be lower than that charged to the first client since Alpha could use the same design probably with minor changes. Alpha is not quite sure how to respond to the second client's request.

With the aid of the Code of Ethics, discuss Alpha's professional obligations under the circumstances.

Question # 4 (20 Marks)

Omega, a professional engineer, is employed on a full-time basis by a well-known radio broadcast equipment manufacturer as an engineering sales representative. In addition, Omega gives professional engineering advice to organizations and practitioners in the radio broadcast field. Typically, Omega would analyze their technical problems and, where required, would make recommendations concerning purchasing and installing certain radio broadcast equipment to rectify the problems identified. In many cases, Omega recommended the use of broadcast equipment manufactured by his / her employer.

(10 Marks)

a) Are there any specific requirements that Omega must satisfy in order to be authorized to engage in the kind of additional professional engineering activity described above? If yes, what are they?

(10 Marks)

b) Under the circumstances, discuss the obligations, if any, that Omega has to his / her employer?

2.2 *Suggestive answers to Ethics, AUGUST 2003*

Answer to question 1

(a)   Each Professional Engineer is provided with a seal indicating that the individual is licensed. Only final drawings, designs, etc must bear seal and signature of the Professional Engineer who has prepared it or has directly supervised it. A Professional Engineer who signs and seals a drawing, etc. that has not been prepared by him/her or under his/her direct supervision may be guilty of professional misconduct. Since this seal has a legal significance, it should not be used casually. In order to avoid any modification, only prints must be signed and sealed. Also, releasing drawings, which bear a photocopy of the Professional Engineer's stamp and signature may be easily duplicated.

(b)   Corporations that offer services to the public must be licensed. This licence, which is called "certificate of authorization" has two main purposes. 1- to identify the persons who are responsible for the engineering work in the corporation. 2- to ensure that their qualifications are acceptable. The requirements for obtaining such a certificate are; i- five years experience in the related field. ii – possession of professional liability insurance (it is voluntary)

(c)   Since a holder of limited licence is not a member of the Association, he/she is not entitled to use the title of P. Eng.

(d)   In order to qualify to be a consulting engineer, the applicant must
  1- have two years experience in private practice
  2- pass examinations assigned by the Association Council or be exempted from it.

(ii)

Association of Professional Engineers regulates profession of engineering. Majority of members of the Association's Council are elected by and from the members of the Association. Association's Regulations and by laws are approved by the Council. This Council also directs the staffs who administer the Professional Engineers Act and Regulations. The disciplinary committee formed by licensed engineers are appointed by the same Council to disciplines Professional Engineers. Since engineers themselves govern the engineering profession, it is, therefore called "self – regulating".

(b) Pursuant to the Code of Ethics, every professional engineer is duty bound to protect public interest in engineering matter. He / she must act as a guardian to safeguard life, properties, health of the general public. In accordance with the Code of Ethics, a professional engineer must at all time behave with honesty and integrity. A professional engineer must always retain his / her impartiality when the engineer is hired as an expert witness. He / she is also bound by the Code of Ethics to protect the environment.

Answer to question 2

This is a special type of conflict of interest. Since you had a partnership with the person who is now an officer in Octopus Enterprises, it will be viewed as though you are doing a favour for your friend, your old partner, if you approve a major engineering contract to Octopus. Although, your decisions may be based on impartiality, but it will not be observed to be fair by the general public. In particular, since it is a government contract and oppositions will soon confront with the truth and therefore expose such conflict of interest.

Pursuant to the Code Ethics, a professional engineer must "avoid or disclose a conflict of interest that might influence the practitioner's action or judgment". So, in order to avoid this type of conflict of interest, you better send all the documents to your superior who does not have a conflict of interest. Failing to disclose a conflict of interest in awarding a major contract is a breach of the Code of Ethics. You may even lose your job.

Answer to question 3

First you should know who owns the copyright for the drawing, Alpha or the client. Now, in order to establish a fair and reasonable fee, you must have clear information about the required knowledge and qualifications. You must also consider difficulty and responsibility as well as the time duration to complete the job.

In the given case, the second client benefits from a design that has already been tested and most probably, less time may be required. But, level of knowledge and responsibility for Alpha will be the same. Consequently, it would not be fair for Alpha to reduce her fees.

Answer to question 4

(a) Omega who is a professional engineer and full-time employee of radio broadcast equipment manufacturing has also involved himself / herself in rendering his / her professional services to organizations and practitioners in radio broadcast field. Such a part time job is known as moonlighting. Pursuant to the Code of Ethics, any professional engineer who moonlights during his / her leisure time may be required to obtain a permit known as "Certificate of Authorization". Omega better obtain liability insurance (though it is not compulsory in Ontario). Omega has also a conflict of interest, because Omega recommends the organizations, where he / she moonlights, to buy the equipments manufactured by his /her employer. According to the Code of Ethics, conflict of interest must be avoided, otherwise it must be disclosed. Pursuant to the Code Ethics, any professional engineer who is involved in conflict of interest but does not disclose it may be charged with professional misconduct.

(b)   The given case indicates that Omega "moonlights". It is not unethical to moonlight. It, obviously, requires extra energy and stamina to work during weekends and late at nights. But, the Code of Ethics requires the employee, who is moonlighting, to be honest and loyal to his / her original employer. Consequently, the employee must not compete with his / her original employer and the employer must be fully informed of the employee's pat-time job. Therefore, Omega has an obligation to fully inform his / her employer.

ASSOCIATION OF PROFESSIONAL ENGINEERS OF ONTARIO

PROFESSIONAL PRACTICE EXAMINARTION-August 9, 2003

PART "B" – Engineering Law and Professional Liability

This examination comes in two parts (**Part "A" and Part "B"**) which must be completed within the allotted **three-hour** period. Both parts of the examination must be passes at the same sitting in order to be assigned a "PASS" for the Professional Practice Examination. The result of the combined examination, Part "A" plus Part "B", will be declared a "PASS" or "FAIL"

Use the correct colour – coded Answer Book for each part, place in the correct envelope and **seal after completed.**

> *White Answer Book for Part A white question paper:*
> *Coloured Answer Book for Part B coloured question paper:*

This is a "**CLOSED BOOK**" examination. **No** aids are permitted other than the excerpts from the 1990 Ontario Regulation 941 covering sections 72 (Professional Misconduct) and 77 (Code of Ethics) supplied at the examination. Dictionaries are **not** permitted.

The marking of the questions will be based not only on academic content, but also on legibility and the ability to express oneself clearly and correctly. If you have any doubt about the meaning of a question, submit with your answer a clear statement of how you have interpreted the question.

All **four** questions constitute a complete paper for Part "B", Question 1 is in two parts totaling 40 marks, including a choice of four definition type questions of equal value[1(a)] and four separate questions of equal value [**1(b)**]. . Questions 2, 3 and 4 are each worth 20 marks.

2.3
(MARKS)

(40)   1.   Please not that this question 1 is a two-part question, worth 40 marks in total.   Question 1(a) is a definition type question, worth 20 marks. Question 1(b) consists of four parts of equal value, totaling 20 marks.

(20)   (a)   <u>Briefly</u> define and explain any <u>four</u> of the following:

        (i)   Liquidated damages
        (ii)   Rule of contra proferentem
        (iii)   Duress
        (iv)   Frustration of contract
        (v)   Gratuitous promise
        (vi)   Director's standard of care

(20)   (b)   (i)   Explain the concept, and application, of equitable estopped.

        (ii)   what are the principles applied to determine the enforceability of a non-competition agreement that restricts an employee from working for a competitor after leaving his or her employment.

        (iii)   Explain the meaning of "limitation period" and provide example of limitation period relevant to engineers and contractors.

        (iv)   List five examples of inappropriate conduct in the workplace under the Ontario Human Rights Code

(10 marks for part (a))   2 (a)   An information technology hardware supplier ("BIDCO") submitted a fixed price bid on a major computer installation project for a large engineering firm in response to the engineering firm's request for proposals. BIDCO included with its tender, as required a certified cheque for $100,000 payable to the engineering firm as a tender deposit.

The request for proposal also provided that if the tender was accepted by the engineering firm and the successful bidder did not execute the contract enclosed with the request for proposal the engineering firm would be entitled to retain the tender deposit for its own use and to accept any other tender.

BIDCO made a clerical error in compiling its tender submission omitting an amount of $ 1,000,000 from its tender price of $ 6,000,000. BIDCO drew the clerical error to the attention of the engineering firm within 5 minutes after the official time foe submitting bid had expired. BIDCO indicated that it wished to

2 of 5 pages   August 9, 2003: Part B - PPE

withdraw its tender but the engineering firm refused to allow it to do so and awarded the supply contract to BIDCO.

Was BIDCO entitled to withdraw its bid? Was the engineering firm entitled to keep the tender deposit? Please provide your reasons and analysis, explaining the relationships and indicate a likely outcome.

(10 marks for part (b))   (b)   An engineer engaged as an environmental consultant ("ENGCO") entered into a contract with an owner ("OWNERCO") of a piece of land to conduct an environmental compliance audit.. OWNERCO was considering selling the land.

ENGCO included in its environmental report the following provision:
" This report was prepared by ENGCO for the account of OWNERCO. The material in it reflects ENGCO's best judgment in light of the information available to it at the time of preparation. Any use which a third party makes of this report, or any reliance on decisions to be made based on it, are the responsibility of such third parties. ENGCO accepts no responsibility for damages, if any, suffered by any third party as a result of decisions made or action based on this report."

If ENGCO's report contained negligent misstatements, could a third party who had subsequently purchased the land from OWNERCO succeed in a tort claim against ENGCO? Explain.

(20)   3.   Provincial Life of Ontario Inc. ("Provincial"), an insurance company, retained an architect, to design a new corporate head office in North York, Ontario. Provincial, as client, and the architect entered into a written client / architect agreement in connection with the project. According to the agreement, the architect was to prepare the complete architectural and engineering design for the project.

In order to carry out the structural engineering aspects of the design, the architect engaged the services of a structural engineering firm. The architect and the structural engineering firm entered into a separate agreement to which Provincial was not a party.

To determine the nature of the soil on which the project would be constructed, two shallow test pits, each about 1.25 meters deep, were dug on the site at locations selected by the architect. The architect

telephoned the structural engineering firm's vice-president and requested that the firm send out a professional engineer to examine the soils exposed in the test pits.

Based on information received from the professional engineer sent to examine the soil, the vice president of the structural engineering firm reported to the architect that the test pits had revealed a silty clay. The vice-president also recommended to the architect that a soils engineer be engaged to carry out more thorough and proper soils tests. The architect rejected the recommendation stating that there was not "enough room in the budget" for more soils test.

The architect succeeded in persuading the vice-president to send a letter to Provincial giving a "soil report" based on the examination of the shallow test pits. The vice-president stated in a letter to Provincial, that based on its examination of the test pits, the soil was a fairly uniform mixture of clay and silt which would be able to support loads up to a maximum of 100 KPa.

The structural engineering firm then completed its structural engineering design on the basis of the maximum soil load reported to Provincial.

The project was constructed in accordance with the plans and specifications. Subsequently, the building suffered extensive structural change, including severely cracked and uneven floors and walls.

On the basis of an independent engineering investigation by an engineer retained by Provincial, it was determined that the extensive structural change in the building had resulted from the substantial and uneven settlement of the building. The investigation also determined that the subsoil in the area of the building consisted of 30 to 40 meters of compressible marine clay covered by a surface layer of dryer and firmer clay two meters in depth. The investigation also revealed that the test pits that were dug had not penetrated the surface layer into the lower layer of compressible material.

What potential liabilities in tort law, arise from the preceding set of facts? Please state the essential principles applicable to a tort action and apply these principles to the facts above. Indicate a likely outcome of the matter.

4 of 5 pages   **August 9, 2003: Part B – PPE**

(20)   4.   Hyper Eutectoid Steel Inc. ("HESI") is a company which produces various types of steel for industrial applications. In order to increase the strength of its steel products, HESI uses a process of quenching and tempering. During the quenching stage, hot steel is quickly cooled with water. During the tempering stage, the steel is then heat treated for an appropriate time. The process requires large amount of water and heat.

Faced with rising cost for energy, HESI decided to install a heat recovery system. The system would include a heat exchanger by which heat could be recovered from the cooling water in the quenching stage, combined with additional heat from a steam line in the plant that was otherwise not being fully utilized. The recovered heat, then, would be used to heat the steel in the tempering stage.

HESI entered into an equipment supply contract with Energy Recovery and Recyclings Systems Inc. ("ERRS"). ERRS agreed to design, supply and install a heat recovery unit for a contract price of $600,000. After an analysis of HESI's processes ERRS determined and guaranteed in the contract that the heat recovery system would recover 40% of the heat in the cooling water and that this would result in substantial savings in energy costs.

The contract also contained a provision limiting ERRS's total liability to $600,000 for any loss, damage or injury resulting from ERRS's performance of its services under the contract.

The heat recovery system was installed and was operational; however, certain defects in the heat exchanger prevented the system from ever recovering more than 5% of the heat in the cooling water. After repeated unsuccessful attempts by ERRS to remedy the defects, HESI terminated its contract with ERRS and hired another supplier, who, for an additional $800,000, replaced the heat exchanger and was able to achieve the level of performance originally promised by ERRS. The total amount that had been received by ERRS under its contract (prior to termination) was $500,000.

Explain and discuss what contract law claim HESI can make against ERRS in the circumstances. In answering, please include a brief summary of the development of relevant case precedents relating to the enforceability of contract provisions that limit liability.

2.4 Suggestive answers to Law, AUGUST 2003

Answer to question 1

(a)

(i)   Liquidated damages or pre- estimated damage is a clause in a contract. This clause pre-estimates damage, which is likely to occur due to non-performance of a contract. It is also the clause by which a contractor limits his liability for tort damages.

(ii)   It is a rule that indicates the importance of a clear and unambiguous language in preparing a contract. This rule emphasizes that a contract must be free from ambiguity. If a contract contains ambiguity then it will be interpreted against the party who poorly drafted it.

(iii)   Duress means you are forced to do it. It is when a party to a contract uses violence, threat to force another person to enter into a contract. The threat must be directed at the person, his/her property or his/her close relatives. This type of contract is not enforceable.

(iv)   An unexpected and highly rapid change of circumstances may cause the party to a contract not to be able to fulfill his/her contractual obligations. When such exceptional circumstances occur, then the contract is frustrated and is discharged by frustration. Occurrence of war is a good example. However, the concept of discharge by frustration cannot be applied to justify discharge of a contract by frustration due to changes of circumstances, which were beyond their contemplation.

(v)   It is a promise without consideration. A promise or a contract without consideration cannot be enforced. An exception exists in the case of equitable estoppel .

(vi)   Every director or officer of a corporation must use his / her power
    a- with honesty and in good faith to protect interest of the corporation
    b- with care, diligence and skill that a prudent person would exercise in that
        circumstances.

Answer to question 1

(b)

  (i)   It is when a party to a contract makes a gratuitous promise, a promise without consideration in such a way that the second party to the contract reasonably relies on it, then it would be unfair and inequitable to allow the first party to revert to his strict contractual terms.

(ii)   Courts always apply the principle of reasonableness to determine if the non-competition agreement is enforceable. The court must be convinced that such contract is not against public policy. Courts are reluctant to enforce restrictive covenants that would limit former employee's ability to earn a livelihood.

(iii)   Limitation period is a specified period of time during which a claim for the damages   can be made. If a claim is made after expiry of the period, then the claim would not succeed. This is known as "statute barred".

(iv)   The inappropriate conducts in the work place as have been prohibited by Ontario Human Rights are: discrimination based on race, place of origin, skin colour, ethnic origin, sex, handicap and age. The Code also outlaws sexual harassment.

Answer to question 2

(a)

There are two types of contraction contracts. Contract A and contract B. Contract A is an irrevocable contract and it must be submitted with a consideration. (In the given case A certified cheque for $ 100,000). When the bidder submits his bid in response to a call for bid, then contract A is formed. If the lowest bidder tries to withdraw his bid, then the owner has to select second bidder (assuming that selection is based on the lowest bidder). The difference in amount between the second lowest bidder and the lowest bidder is the damage that the owner would suffer. This is called the direct damage. When a bidder realizes that there has been an error or mistake in the bid, he is obliged to persuade the court that the error is inconsistent with the term actually agreed upon. This error must be of secretarial or recording nature.

In the given case, BIDCO submitted his bid that means contract A was formed. BIDCO also drew attention of the firm to the error. Consequently, it would be unjust to let the owner enrich himself for the mistake of BIDCO. Since there exists a strong possibility that court may issue the order of rectification therefore, the owner would be entitled to retain the cheque and even if the work went on, it will be a poorly performed job and with a sense of grievance.

(b) Courts, in many cases, have absolved the party who has produced a statement disclaiming responsibility to the third party from any liability. Since in the given case, it is clearly written that ENCO accepts no responsibility for damages, if any, suffered by any third party, (disclaiming the responsibility) then the land purchaser does not have any legal rights to take any action against ENCO.

Answer to question 3

The principle of tort, which means injury, is to compensate a person who has sustained injury as a result of negligent conduct of the other person. In order to be successful in tort claim against the second person, there is no need for a contract to exist between them. The plaintiff must prove that:
i-   the defendant owed the plaintiff a duty of care

      ii-   the defendant breached that duty by his conduct
      iii-   the defendant negligent conduct caused injury to the plaintiff

The architect was in direct contract with the Provincial. He knew or ought to have known that Provincial was relying on his professional conduct. He owed the Provincial a duty of care. He should have foreseen the consequences that may arise as result of his careless performance. He persuaded the vice-president to send the letter, while he knew that the report (letter) did not contain accurate information and more testing should have been done. So he breached that duty by his negligent conduct. As a result the building suffered extensive structural damages, the plaintiff sustained damages.

Since the vice- president misrepresented the information, and since it is a tort case, he should also be liable to the damages. This is known as "concurrent tortfeasors". But since the vice president is an employee of a firm, his firm may be vicariously liable to the damages.

Answer to question 4

This case is dealing with fundamental breach of contract. The contract contains an exemption clause, limiting liability of ERRS to $ 600,000 for any loss, damage or injury.

At one time Canadian courts followed English courts precedent in applying what was called the "fundamental breach doctrine". According to this doctrine, an exemption clause cannot be enforced in case of a fundamental breach. Eventually, this doctrine was overruled by English courts. Even though the concept of this doctrine is not completely overruled by Canadian courts, but strong preferences are given to the exemption clauses in commercial cases. Canadian courts following English courts do not disregard the exemption clauses.

In the given case, the contract is for $ 600,000. HESI has spent $500,000+800,000=1300,000 dollars to get the job done. ERRS has limited its total liability to $600,000. Since Canadian courts do not disregard the exemption clause, therefore HESI would be entitled to receive only $ 600,000 from ERRS.

CHAPTER 3

PPE, Part "A", Part "B", December 2003

Abstract. This chapter discusses in a very informal manner the suggestive answers of the author to the question papers of December 2003. The two parts (Part "A" – Professional Practice and Ethics) and (Part "B" – Engineering Law and Professional Liability) of the three Professional Practice Examinations have been completely answered. The simplicity of the language adopted in this chapter will act as a useful tool for the PPE candidates to further understand these two subjects.

ASSOCIATION OF PROFESSIONAL ENGINEERS OF ONTARIO

PROFESSIONAL PRACTICE EXAMINARTION-December 6, 2003

PART "A" – Professional Practice and Ethics

This examination comes in two parts **(Part "A" and Part "B")**. Both parts must be completed in this sitting. You
will be given a total of **180 minutes** to complete the examination.

Use the correct colour – coded Answer Book for each part, place in the correct envelope and **seal after completed.**

> *White Answer Book for Part A white question paper:*
> *Coloured Answer Book for Part B coloured question paper:*

This is a **"CLOSED BOOK"** examination. **No** aids are permitted other than the excerpts from the 1990 Ontario Regulation 941 covering sections 72 (*Professional Misconduct)* and 77 (*Code of Ethics*) supplied at the examination. Dictionaries are **not** permitted.

The marking of the questions will be based not only on academic content, but also on legibility and the ability to express yourself clearly and correctly in the English language. If you have any doubt about the meaning of a question, please state clearly how you have interpreted the question.

All **four** questions constitute a complete paper for Part "A". Each of the four questions is worth 25 marks.

Where a question asks if a certain action by an engineer was ethical or not, a simple "yes" or "no" answer is not sufficient. You should identify the applicable clauses in Regulation 941 and comment on the action in each situation.

Part A – PPE, December 6, 2003   Page 1 of 4

3.1

Question 1

(10)　　(a)　Two of the PEO's functions are discipline and enforcement. Explain what enforcement is and how it differs from discipline and name two specific activities that are subject to enforcement.

(10)　　(b)　recently, the Professional Engineers Act and its Regulations were amended to add a new type of licence called the Provisional Licence. What is the purpose of the Provisional Licence and how is a Provisional Licence different from the "full" Licence given to a Member of the PEO?

(5)　　©　Is it mandatory that a holder of a Certificate of Authorization maintain insurance?

Question 2

Egoist is a P. Eng. employed by EngCo, an engineering company. As Chief Project Engineer, Egoist is in charge of a project for BigBucks, an important client of EngCo. BigBucks and Egoist have several disagreements over design that Egoist has developed. BigBucks wants a cheaper, more conventional solution. Egoist is convinced that the design is a "masterpiece" and believes that BigBucks "doesn't" have an ounce of imagination". Egoist simply shrugs off BigBucks and refuses to discuss any other alternative.

BigBucks is furious and phones Sly, the President of EngCo, to yell and complain about Egoist. BigBucks threatens to hire another engineering firm to complete the design according to BigBucks' wishes.

You work for EngCo as an intermediate design engineer. Sly calls you into a private office and closes the door. Sly asks you to review Egoist's design and instructs you to keep the review a secret from Egoist. Sly explains that Egoist is a senior engineer who has been with EngCo for 28 years and could be "a bit sensitive at times".

(15)　(a)　What do you tell Sly?
(10)　(b)　Please comment on Egoist's conduct in dealing with BigBucks. How should Egoist have responded to BigBucks' request?

Question 3

AgriFab is a designer and manufacturer of farming equipment.

Recently, Farmer was seriously injured while operating a tractor designed and manufactured by AgriFab. In a letter to AgriFab, Farmer's lawyer claimed that the injury was due to a malfunction caused by a design error by AgriFab's engineering department. The letter threatened that AgriFab would be sued on account of Farmer's injuries.

AgriFab retains you (a Consulting Engineer) as an expert. Your services would be to investigate the failure and to give AgriFab your expert opinion on the cause of the failure. If the case goes to court, you could be called to testify as AgriFab's expert witness. For your services, AgriFab would pay you at an hourly rate. If you are called to testify in court and AgriFab wins the case, AgriFab would pay you a bonus in addition to your hourly rate.

Following your investigation, you conclude that the tractor was not designed properly and that Farmer was injured when certain safety features of the tractor failed to function. You also conclude that it is likely that other farmers could be seriously injured while operating the particular tractor model.
You report your conclusions to AgriFab.

Based on your report, AgriFab promptly agrees to pay Framer $1 million. In exchange for the payment, Farmer agreed to give up the lawsuit and agreed to keep the payment a secret. The secrecy agreement was very important to AgriFab because AgriFab did not want future tractor sales to suffer from bad publicity.

AgriFab thanks you for your services and pays your fee.

(20)   (a)   Is there anything else you should do?

(5)   (b)   Please comment on the appropriateness of the fee structure according to which you would be paid.

Question 4

WorldEng, a large engineering firm, was hired to prepare the design for a chemical production plant for MegaChem. In addition to preparing the plant design, WorldEng's duties included providing inspection services during the construction stage of the project. The project was completed successfully.

You are a P. Eng. and have been employed on a full-time basis by WorldEng for several years. You work in the Process Division and are involved on several process design projects. You were an important member of the design team that prepared the design for MegaChem's plant. In addition to working for WorldEng, you supplement your income by occasionally undertaking work on weekends and during evening for EngInc, another engineering company. A colleague of yours, who is a P. Eng. at EngInc, assigns you such work and assumes responsibility for it.

A few years after the plant was completed, MegaChem decided to restructure its operations and sell the plant. BuyerCo has agreed to buy the plant, but before it does so, BuyerCo wants to satisfy itself (and its bank) that the plant was built to proper standards and is in good physical condition. BuyerCo hires EngInc. to inspect the physical plant and to review relevant documents (including the original plans and specifications, "as-built" drawings, and operations and maintenance logs). EngInc is very busy on several projects and asks you to assist with the plant inspection and document review.

(10)   (a)   Comment on the appropriateness of your employment arrangements.

(10)   (b)   assuming that your employment arrangements have not changed since the plant was designed and constructed, how do you respond to EngInc's request for assistance?

(5)   ©   Would you need a Certificate of Authorization to provide services to EngInc.?

3.2 Suggestive answers to Ethics, DECEMBER 2003

Answer to question 1

(a)

Enforcement: When a person who is not licensed by the PEO infringes the Professional Engineers act and practices engineering without licence or uses an engineering seal and the title "Professional Engineer" or any title that may make people believe that he/she is really a Professional Engineer, has violated the Professional Engineering Act. This person will be prosecuted, in accordance with the Act, in the court. If convicted, he/she may have to pay severe fines. Whereas, when a licensed engineer violates the Professional Engineering Acts or Regulation 941, such as professional misconduct, then the Engineer will be disciplined by the PEO Discipline Committee. This Committee is authorized to revoke or suspend the licence or certificate of authorization of the individual. This Committee may also impose restrictions or fine.

(b) Provisional Licence: The Registrar of the PEO may grant a provisional licence, which is valid for 12 months to an applicant who complies with the requirements. It may be renewed if the Registrar believes that renewal is necessary. Holder of provisional licence cannot sign and seal the final documentation unless the documentation is first signed and sealed by the professional who has supervised the work. This seal includes the holder's surname, the word "Provision Licensee" and Association of Professional Engineering of Ontario, as well as licence number and the date of expiry. The licence given to a member of the Association, known as P. Eng., authorizes him/her to supervise and direct those who are working under his/her supervision. A P. Eng. has full authority to affix his/her seal and signature on the final documentations and admitted to practice professional engineering all over Ontario.

© No, it is not mandatory that a holder of a Certificate of Authorization should maintain liability insurance. In such a case the engineer must inform his/her client in writing that the engineering services are not covered by liability insurance. The client must acknowledge and accept the condition.

Answer to question 2

(a) Evaluating performances of employees is a common practice and it is the engineering manager who conducts such evaluation. But, the manager should not ask an engineer to review the work of another engineer without his knowledge. A secret review is like a trial in absentia. You must tell Sly that even though, there is no need of having permission to review Egoist' design, but Egoist must be informed prior to reviewing the design. In fact, reviewing Egoist's design for competence and accuracy is a good practice, but Egoist must be informed. It is extremely advisable to check calculations and design of an engineer, but such a review must not be done in secret.

(b) Egoist is an employee and BigBucks is a client. Egoist, as a professional engineer is duty bound to consider welfare and safety of the general public as paramount. Since

BigBucks is not a professional engineer he is not qualified to make any judgment regarding the design.

But since it is his budget, he is qualified to forward his suggestion for a cheaper design with respect to the reality that safety of the general public has been considered as paramount. There is a strong possibility that even we as professional engineers may commit unprofessional acts that we may later regret. When a professional engineer is working for a client, he/she must be courteous and fair. Egoist should have not shrugged off BigBucks. She should have discussed the matter in more details with BigBucks.

Answer to question 3

(a) You are a professional engineer and under every Code of Ethics, your paramount responsibility is to ensure the welfare and safety of the general public. An outside court settlement is something that is commonly done. But in the given case, such settlement indicates that there would not be any changes and/ or modifications to the design of the tractor. Consequently, safety of the public and in particular the framers who will be utilizing AgriFab's designed tractors are in danger. You, as a professional engineer must insist on eradication of unsafe parts. The agreement was very important to AgriFab because AgriFab did not want future tractor sale to suffer from bad publicity. But, it is not something that a professional engineer should agree with. If AgriFab is not welling to comply with your instruction, you must immediately get help from the concerned authorities. You may even blow the whistle.

(b) An expert witness is expected to express his/ her opinion. Therefore, you as a professional engineer must maintain your impartiality towards the outcome of the case. However, you as a recipient of bonus would have a conflict of interest. It is unethical to accept bonus. You must bill AgriFab for your time.

Answer to question 4

(a) You are an employee of WorldEng, but you undertake work on weekends and during evenings for EngInc. Such part-time employment is known as "moonlight". In fact, it is ethical to work for more than one employer, but it requires determination and stamina. You are required by Code of Ethics that while you are moonlighting, you must disclose the situation to the original employer. In addition to this, you should not compete with the original employer and also, moonlighting should not result in reduction of your efficiency during your usual work hours for the original employer. Since you have not informed your original employer about your part-time job, then your conduct is obviously unethical. Such disclosure enables your original employer to judge if you are competing with his work.

(b) This is a special type of conflict of interest. You are a full time employee of WoldEng. You were an important member of the design team that prepared the design for MegaChem. It will be judged as though you are doing a favour for yourself, if you assist with the plant inspection and document review. Your decisions may be based on impartiality, but it will not be observed to be fair by the general public. Once the truth is revealed, you will be exposed to conflict of interest.

Pursuant to the Code Ethics, a professional engineer must "avoid or disclose a conflict of interest that might influence the practitioner's action or judgment". So, in order to avoid this type of conflict of interest, you better inform EngInc of the fact that you are a full time employee of WorldEng and you participated in preparation of the design for MegaChem's plant. You better ask EngInc to send another person who does not have a conflict of interest. Failing to disclose a conflict of interest in awarding a major contract is a breach of the Code of Ethics.

© Certificate of Authorization is required when a person is offering a service directly to the public. In the give case, your colleague is a P. Eng. at EngInc and he / she resumes responsibility for your work. Consequently, you do not need a Certificated of Authorization.

ASSOCIATION OF PROFESSIONAL ENGINEERS OF ONTARIO

PROFESSIONAL PRACTICE EXAMINARTION-December 6, 2003

PART "B" – Engineering Law and Professional Liability

This examination comes in two parts **(Part "A" and Part "B")**. <u>Both</u> parts must be completed in this sitting. You will be given a total of **180 minutes** to complete the examination.

Use the correct colour – coded Answer Book for each part, place in the correct envelope and seal after completed.

> *White Answer Book for Part A white question paper:*
> *Coloured Answer Book for Part B coloured question paper:*

This is a "**CLOSED BOOK**" examination. **No** aids are permitted other than the excerpts from the 1990 Ontario Regulation 941 covering sections 72 (*Professional Misconduct*) and 77 (*Code of Ethics*) supplied at the examination. Dictionaries are not permitted.

The marking of the questions will be based not only on academic content, but also on legibility and the ability to express yourself clearly and correctly in the English language. If you have any doubt about the meaning of a question, please state clearly how you have interpreted the question.

All **four** questions constitute a complete paper for Part "B". Each of the four questions is worth 25 marks.

3.3

(MARKS)

(25)   1.   <u>Briefly</u> define and explain any <u>five</u> of the following :
- (i)   Secret Commission
- (ii)   Undue influence
- (iii)   Five examples of inappropriate conduct in the workplace (list only)
- (iv)   Rule of contra proferentem
- (v)   Limitation period in tort law
- (vi)   Director's standard of care
- (vii)   Contract
- (viii)   Consequential damages

(25)   2.   An information technology firm submitted a bid to design software and hardware for an electronic technology process to control the operation of a large scale handling and related security facility for a major airline.

The firm's fixed guaranteed maximum price was the lowest bid and the contract was awarded to it. The contract conditions entitled the information technology firm to terminate the contract if the airline did not pay monthly progress payments within 15 days following certification that a progress payment was due. Pursuant to the contract, an independent engineering firm engaged as contract administrator carried out the certification.

The work under the contract was to be performed over an 8 month period. After commencing work on the project the information technology firm determined that it had made significant judgment errors in arriving at its bid price and that it would face a major loss on the project. Its concern about the anticipated loss was increased further when it also leaned that, in comparison with the other bidders, its bid price was extremely low and that, in winning the bid, by comparison with the other bidders, it had left more than two million dollars "on the table".

Three monthly progress payments were certified as due by the independent engineering firm and paid by the airline in accordance with the terms of the contract. However, after the fourth monthly progress payment was certified as due by the independent engineering firm, the airline's finance department asked the information technology firm's representative on the project for additional information relating to an invoice from a subcontractor to the information technology firm. The subcontractor's invoice comprised a portion of the fourth progress payment amount. The airline's finance department requested that the additional information be provided prior to payment of the fourth progress payment.

There was nothing in the signed contract between the information technology firm and the airline that obligated the information technology firm to provide the additional information on the invoice from its subcontractor. However, The information technology firm's representative did verbally indicate to the airline's finance department that the additional information would be provided.

The information technology firm never provided the additional information relating to the subcontractor's invoice.

Sixteen days after the fourth progress payment had been certified for payment, the information technology firm notified the airline in writing that it was terminating the contract because the airline was in default of its obligations to make payments within fifteen days pursuant to the express wording of the contract.

Was information technology firm entitled to terminate the contract in these circumstances? In giving reasons for your answer, identify and explain the relevant legal principle, its purpose, how it arises, and how it would apply to the facts.

(25)   3.   National Stores Inc. ("NATIONAL"), the owner of a grocery store chain in Ontario, contracted with an architect to design and prepare the construction documentation for a new store in a town in northern Ontario.

The architect produced some general construction specifications that included a requirement that an automatic sprinkler system, conforming to the National Fire Protection Association ("NFPA") standards, be installed.

The architect retained an engineering firm pursuant to a separate agreement to which NATIONAL was not a party. Under the contract the engineering firm was to prepare the detailed engineering design for the project, including the sprinkler system. The engineering design was to conform to the architect's general specifications.

A recent engineering graduate employed by the engineering firm prepared the design of the sprinkler system. Not being familiar with the NFPA requirements, the employee read certain sections of the standards but did not have enough time, given other project responsibilities, to pay close attention to all the details. A professional engineer reviewed the employee's completed sprinkler system design. Although the professional engineer did not perform a detailed check, the professional engineer considered the design satisfactory.

3 of 5 pages   December 6, 2003. Part B-PPE

Six months after the store opened for business, a fire occurred early one morning. The fire caused substantial damage to the store and to its inventory and NATIONAL had to close the store for repair.

NATIONAL retained a consulting engineer to conduct an independent investigation. The consulting engineer determined that the sprinkler system was inadequately designed. Specifically, the design did not conform to the NFPA standards, which required, among other thing that the coverage per sprinkler head was not to exceed 10 square meters. The engineer determined that 10 percent of the sprinkler heads were designed to cover an area as high as 25 square meters. The report indicated that, in the engineer's expert opinion, had the sprinkler head spacing conformed to the NFPA standards, the fire should have been quickly extinguished and would not have spread to any great extent.

What liabilities in *tort law* may arise in this case? In your answer, explain the purpose of tort law and identify what essential principles of tort law are relevant. Apply each principle to the facts. Indicate a likely outcome of the matter.

(25)   4.   ACE Construction Inc. is a company primarily engaged in the business of supplying heavy equipment used in construction. As part of the company's economic plan to expand its business, ACE became interested in the rock crushing industry.

ACE had become aware that International Metals Company Ltd. ("IMCO") required a contractor to crush, weigh and stockpile approximately 250,000 tons of ore. As ACE believed this was an excellent opportunity to venture into the rock crushing business, it decided to tender on the IMCO contract.

In order to tender on the contract, ACE set out to purchase the necessary equipment to crush the material. ACE was contacted by a representative of Rock Busters Ltd., a company which sold such equipment. After visiting the IMCO site and determining the nature of the material to be crushed, the representative discussed the IMCO contract with ACE. After performing a number of calculations, the representative determined and guaranteed that the equipment Rock Busters would provide would be capable of crushing the materials at a rate of 175 tons per hour. On the basis of the guarantee, Rock Busters and ACE entered into a contract. Rock Busters agreed that if ACE were successful in its tender to IMCO, Rock Busters would provide the equipment for a price of $400,000. The contract also contained a provision limiting Rock Busters' total liability to $400,000 for any loss, damage or injury resulting from Rock Busters' performance of its services under the contract.

Based on the information provided by the representative, ACE prepared and submitted its tender to IMCO. IMCO accepted the tender and entered into a contract with ACE to crush the material.

The rock crushing equipment was set up at the IMCO site by employees of Rock Busters and crushing operations commenced. However, from the beginning there was trouble with the operation. One of the components of the crusher, called the cone crusher, consistently became plugged by the accumulation of material. Each time the cone crusher became plugged, the operation would have to be shut down and the blockage cleared manually. In some cases, such blockages caused damage to the equipment. Rock Busters made several unsuccessful attempts to correct the defect by making modifications at the site and at its factory. The crushing equipment was never able to crush more than 30 tons of materials per hour.

In order to meet its obligations under the IMCO contract, ACE hired another supplier to correct the defects in the Rock Busters equipment. For an additional $500,000 the supplier replaced the cone crush with one manufactured by another company. The modified equipment was able to crush the material at the rate of 180 tons per hour. The total amount which had been paid by ACE to Rock Busters was $350,000.

Explain and discuss what claim ACE can make against Rock Busters in the circumstances. Would ACE be successful in its claim? Why? In answering, please include a summary of the development of relevant case precedents. In particular, point out how the law changed because of these relevant case precedents. Identify the legal principles on which the precedent decisions were based and apply the appropriate legal principle to the facts.

3.4 Suggestive answers to Law, DECEMBER 2003

Answer to question 1

(i)   It is money, or something of value that has been offered through a secret agreement to show a favour or disfavour to a third party. Secret commission is an offence under Criminal Code of Canada. If a Professional Engineer is involved in secret commission, then he/she will be charged under Regulation 941/90.

(ii)   It is similar to duress. It is when a person dominates the free will of another person to the extent to induce him/her to an unfair contract. Such as, relation between wife and husband, parent and a young child. Such a contract is not enforceable, and it is therefore void.

(iii) The inappropriate conducts in the work place as have been prohibited by Ontario Human Rights are: discrimination based on race, place of origin, skin colour, ethnic origin, sex, handicap and age. The Code also outlaws sexual harassment.

(iv)   It is a rule that indicates the importance of a clear and unambiguous language in preparing a contract. This rule emphasizes that a contract must be free from ambiguity. If a contract contains ambiguity then it will be interpreted against the party who poorly drafted it.

(v)   Limitation period is a specified period of time during which a claim for the damages can be made. If a claim is made after expiry of the period, then the claim would not succeed. This is known as "statute barred".

(i)   Every director or officer of a corporation must use his / her power
         a- with honesty and in good faith to protect interest of the corporation
         b- with care, diligence and skill that a prudent person would exercise in that circumstances.

(vii)   In order for a contract to be enforceable, it must contain the five essential terms.
         a- an offer made and accepted
         b- mutual intent to the contract
         c- consideration
         d- capacity to contract
         e- lawful purpose
         Only valid contracts are enforceable.

(viii) Consequential damages are also known as indirect damages. When a contractor is working on an owner property and due to his negligent services the adjacent land or firm sustains damages, then indirect or consequential damages have occurred. There is a possibility that an owner may be fined due to the contractor's non-compliance with the environmental protection.

Answer to question 2

This is the case, which involves "equitable estoppel", which states that when a party to a contract makes a gratuitous promise, a promise without consideration, and the second party to the contract reasonably relies on it, then it would be unfair and inequitable to allow the first party to revert to his contractual terms. Court always enforces only those terms and conditions that are clearly and unambiguously written in a contract and a contract without consideration cannot be enforced. If a term has verbally been agreed upon, but has not been entered into the contract, then this term is not enforceable. But in case of a gratuitous promise, since it would be unfair and inequitable to let the first party return to his strict contractual term, the concept of equitable estoppel is well applicable.

In the given case, a representative of the information technology firm made a gratuitous promise, when the representative verbally indicated to the airline's finance department that additional information would be provided. In this case the concept of equitable estopple is well applicable and information technology firm must be estopped. It would be unfair and inequitable to allow information technology firm to return to his strict contractual term.

Answer to question 3

Tort means damage, injury and its main purpose is to compensate a person who has sustained damages as a result of negligent conduct of another person. Claim for tort may arise, even though the two parties are not in contractual relationship. In order for a claim for tort to be successful the plaintiff must prove that:
a-the defendant owed the plaintiff a duty of care
b-the defendant breached that duty by his negligent conduct
c-the defendant conduct caused damages to the plaintiff.
In the given case, the plaintiff is NATIONAL and the defendants would likely be the young engineer and the Professional Engineer, but ultimately it would be their engineering firm.

A recent engineering graduate was assigned to design the sprinkler system. All the specifications were provided to him by the architect. The young engineer knew or ought to have known that NATIONAL would rely on his performance. He owed NATIONAL to perform his job with diligence, skill and care. But, he did not pay close attention to all the details, which means he neglected his duty of care. NATIONAL sustained damages as a result of his negligent conduct. The prerequisite for claim for tort may arise when damage manifests itself. In this case, fire and economic loss associated with closing the store. Professional Engineer is one who must consider the safety and welfare of general public as paramount. He is duty bound. He cannot approve anything without conducting a thorough testing and checking. The Professional Engineer should have foreseen the damages that may arise as a result of his negligent conduct. The Professional Engineer breached that conduct by considering the design to be satisfactory, though he had not performed a detailed check. Since it is a tort case, subsequently the young engineer and the Professional Engineer would be "concurrent tortfeasors". Since they are employees of an engineering firm, then the firm would be vicariously liable for the damages.

Answer to question 4

This case is dealing with fundamental breach of contract. The contract between ACE and Rock Busters is for $ 400,000. The contract also includes an exemption clause limiting total liability of Rock Busters to $ 400,000.

At one time Canadian courts followed English courts precedent in applying what was called the "fundamental breach doctrine". According to this doctrine, an exemption clause cannot be enforced in case of a fundamental breach. Eventually, this doctrine was overruled by English courts. Even though, the concept of this doctrine is not completely overruled by Canadian courts, but strong preferences are given to the exemption clauses in commercial cases. Canadian courts following English courts do not disregard the exemption clauses. In fact parties to the contract are free to select and choose their own terms and conditions. Provided they follow the five essential terms making a contract enforceable.

In the given case, ACE has spent $ 500,000 + 350,000 = 850,000 dollars to get the job done. Rock Busters has limited its total liability to $ 400,000. Since Canadian courts do not disregard the exemption clause, therefore ACE would be entitled to receive only $ 400,000 from Rock Busters.

CHAPTER 4

PPE, Part "A", Part "B", April 2004

Abstract. Complete Answers to the questions of question papers of Part "A" - Professional Practice and Ethics, and Part "B" – Engineering Law and Professional Liability of April 2004 are provided in this chapter. The simple and informal languages used in this chapter will surely pave a path towards easy understanding of these two subjects.

ASSOCIATION OF PROFESSIONAL ENGINEERS OF ONTARIO

PROFESSIONAL PRACTICE EXAMINARTION- April 17, 2004

PART "A" – Professional Practice and Ethics

This examination comes in two parts **(Part "A" and Part "B")**. <u>Both </u>parts must be completed in this sitting. You will be given a total of **180 minutes** to complete the examination.

Use the correct colour – coded Answer Book for each part, place in the correct envelope and **seal after completed.**

> *White Answer Book for Part A white question paper:*
> *Coloured Answer Book for Part B coloured question paper:*

This is a "**CLOSED BOOK**" examination. **No** aids are permitted other than the excerpts from the 1990 Ontario Regulation 941 covering sections 72 (*Professional Misconduct*) and 77 (*Code of Ethics*) supplied at the examination. Dictionaries are **not** permitted.

The marking of the questions will be based not only on academic content, but also on legibility and the ability to express yourself clearly and correctly in the English language. If you have any doubt about the meaning of a question, please state clearly how you have interpreted the question.

All **four** questions constitute a complete paper for Part "A". Each of the four questions is worth 25 marks.

Where a question asks if a certain action by an engineer was ethical or not, a simple "yes" or "no" answer is not sufficient. You should identify the applicable clauses in Regulation 941 and comment on the action in each situation.

Part A – PPE, April 17, 2004   Page 1 of 4

Any similarity in the question to actual person or circumstances is coincidental.

4.1
Question 1

(10) (a) Section 12(1)of the Professional Engineers Act states that no person shall engage in the practice of professional engineering or hold himself, herself or itself out as engaging in the practice of professional engineering without holding one of four specified types of licences. Name each of those four (4) types of licences and state which of those licences indicate that its holder is a "professional engineer", according to the Act.

(5) (b) PEO's Discipline Committee has the power to revoke or suspend any of the licences referred to in Question 1 (a), if the licence holder commits professional misconduct. <u>Besides revocation and suspension,</u> describe three(3) <u>other</u> penalties or sanctions that the discipline committee may impose.

(5) © Is holding one of the above licences sufficient to entitle the holder to offer professional engineering services to the public? Explain.

(5) (d) Are there any restrictions on how professional engineering services may be advertised? Explain.

(25)   **Question 2**

Multicommon Tires Ltd. ("Multicommon") designs and manufactures automobile tires. Recently, Multicommon developed a new polymer which would greatly improve the expected life of its tires. Multicommon has not yet started using the new polymer in its tires. The development of the technology is still in its early stages and the company has not yet obtained a patent.

Multicomon has retained Chem Engineering Inc. ("Chem") to help Multicommon develop a process for putting the new polymer into lage scale production. Chem assigned one of its professional engineers B.N. Evolent, to Multicommon's project. B.N. Evolent is also a member of several trade and professional associations including the International Building Materials Institute (the "IBMI"), an association of designers, manufacturers, sellers and users of building materials. IBMI's mission statement is to improve the building materials industry by the mutual cooperation of its members. B.N. Evolent often volunteers to serve on IBMI's committees.

B.N. Evolent is very intrigued about the new technology and soon realized that the new polymer could improve the durability of building materials made of synthetic rubber. At a recent committee meeting of the IBMI, B.N. Evolent suggested to a group of materials designers that they use the new polymer in their products. B.N.Evolent was happy to share the information and remembered

Any similarity in the question to actual person or circumstances is coincidental.

reading something in PEO's Code of Ethics about practitioners being required to "extend the effectiveness of the profession through the interchange of engineering information and experience". In addition, B.N.Evolent believed, it would be good for the environment and the public interest in general to have more durable building materials" and "besides, Multicommon doesn't make building materials anyway".

Do you agree with B.N. Evolent's conclusions? What do you think of B.N. Evolent's conduct? Explain.

(25)   Question 3

SmallBox Corp. operates a small chain of three retail stores that specialize in selling lumber and other home improvement products. In order to improve its inventory and distribution efficiencies, SmallBox would like to build a central warehouse that would serve all of its stores. SmallBox contacted Engco, a large engineering firm, to inquire about hiring them to design the facility.

Eager is employed as a professional engineer by Engco. At the request of Honcho, the head of Eager's division, Eager accompanied Honcho to a meeting at Engco's offices with some representatives of SmallBox to discuss how Engco might be able to assist SmallBox with the potential project. At the meeting, SmallBox's representatives described to Honcho and Eager the attributes that SmallBox was looking for in the proposed new warehouse. They also asked about the fees that Engco proposed to charge for its services. Upon being advised of Engco's standard rates, SmallBox's representatives stated that, unfortunately, they could not afford to hire Engco for this project. Honcho was not prepared to discount Engco's quoted rates, which Honcho described as being "extremely competitive". Although everyone was disappointed, the meeting ended pleasantly, and both sides politely thanked each other for attending.

The next day, Eager received a telephone call from Frugal, one of the representatives of SmallBox. Frugal was wondering if Eager would be interested in preparing the design for the warehouse "on the side", after work in the evenings and on weekends. SmallBox was prepared to pay Eager at an hourly rate that was 50% of the hourly rate that Engco would have charged for Eager's time. In a hushed voice, Eager undertook to give the proposal some consideration and get back to Frugal.

Eager thought about Frugal's offer. It had been three long years since Eager had last received a salary increase from Engco. The extra money would be nice. Even at rates discounted by 50% from those charged by Engco, this would be a very profitable opportunity for Eager. The money Eager would earn from SmallBox would be more, on an hourly basis, than the rate on which Eager's current salary was based, and unlike Engco, Eager didn't have to worry about big overheads and other expenses. Eager then thought about how Engco might react to the arrangement, but decided that since Engco wouldn't be getting this work anyway, there shouldn't be a problem. Besides, Eager thought, there was no reason why they even needed to know about it.

Eager called Frugal back the next day to accept the engagement and enthusiastically began working on the project that evening.

Please comment on the appropriateness of Eager's conduct.

Any similarity in the question to actual person or circumstances is coincidental.

(25)   Question 4

Elixir Engineering Ltd. ("Elixir") was engaged by Potion Pharmaceuticals Inc. ("Potion") to design a new production line at its pharmaceutical plant in Littletown, Ontario. In addition to preparing the design, Elixir would provide professional services during the installation of the project.

You are a professional engineer employed by Elixir The president of Elixir (who is also a professional engineer) has been very impressed with your work over the years and has personally requested that you be assigned to Potion's project to act as project manager. This is a very important job for Elixir and the assignment represents an important career opportunity for you (rumours have it that a successful project would result in your being promoted to vice president)

The completed design of the new production line includes a sophisticated new proprietary quality control system (designed and manufactured by Calibre Systems Corp. ("Calibre")) The system employs laser coding and machine vision to accurately identify, and label, injection bottles containing various drugs that will be produced at the new facility. The technology uses a laser to code the cap of each injection bottle immediately after it is filled. After the bottles are filled, sterilized and transported through the system, a scanner reads the laser codes on the bottle caps and the appropriate labels are automatically printed and applied to the bottles. With the system, wrongly labeled bottles would be eliminated.

The project has progressed well(both on schedule and on budget) and is ready for start up, commissioning and testing. Unfortunately, during testing, the quality control system malfunctions and incorrectly labels 20% of the bottles. Calibre's technical personnel at the site are able to fix a number of problems contributing to the malfunction, and the error rate is reduced to 3%. Calibre advises you that it will be unable to eliminate the remaining error without redesigning, manufacturing and testing a scanning component, a process which is expected to take at least two months.

You e-mail the bad news about the delay to Potion. The next day, Potion's project representative sends you an e-mail instruction that the production line be accepted with the 3% error rate. Potion's project representative explains that Potion in planning to use the new production line to manufacture a new patented drug, which will benefit thousands of people. Given the immediate demand for the drug, a two-month delay is not acceptable. The representative also indicates that Potion's and Elixir's presidents have already discuss the problem and have agreed that Potion would assume responsibility by legally indemnifying Elixir for any liability that could arise from the incorrect labeling.

What do you do? What do you think of the conduct of Elixir's president? Explain.

Part A – PPE, April 17, 2004   Page 4 of 4

4.2 Suggestive answers to Ethics, APRIL   2004

Answer to question 1

(a)   The four (4) types of licences are: Licence, Temporary Licence, Limited Licence and Provisional Licence.

Pursuant to the Act, a licence permits the member or "licensee" to practice professional engineering all over the Province of Ontario. A temporary or non-resident licence is offered to those who are member of an Association in another province, qualified to work on a specified project and familiar with code and standard of Ontario. Limited Licence, which is rarely issued, permits the holder of the licence to offer his / her engineering services to a specific employer on a specific project. If the holder of the licence leaves the employer, he / she must return the licence to PEO. Provisional Licence: The Registrar of the PEO may grant a provisional licence, which is valid for 12 months to an applicant who complies with the requirements. It may be renewed if the Registrar believes that renewal is necessary. Holder of provisional licence cannot sign and seal the final documentation unless the documentation is first signed and sealed by the professional who has supervised the work. This seal includes the holder's surname, the word "Provision Licensee" and Association of Professional Engineering of Ontario, as well as licence number and the date of expiry.

(b)   The penalties assigned by the Disciplinary Committee are fines, and payment of cost but not imprisonment. In addition to the two penalties mentioned here, the Committee can
   1- impose restriction on the licence
   2- require the member to be reprimanded
   3- require the member to take a course of study.
if a member or a licensee is found guilty. The type of penalty depends on the seriousness of the case.

©   Only members of the Association and holders of temporary licence are permitted to offer professional services to the public.

(d) According to the regulations made under the authority of the provincial Act, advertising is permitted in Ontario, but it should be done in a professional manner. It should be factual. It should not exaggerate. The advertisement should not criticize other person(s) in anyway. Use of engineer's seal and seal of the Association is strictly prohibited.

Answer to question 2

 Under the Code of Ethics, professional engineers are required to maintain confidentiality of his / her client's affairs. It is also possible that a professional engineer may be asked by his/her client to sign a confidentiality agreement. In the given case, even though professional engineer, B. N. Evolent, has not been asked to sign such an agreement, but according to the Code of Ethics it is his duty to keep his client's affairs secret. Therefore, B. N.Evolent should not have disclosed information about the new polymer to material designers. Information and experience that an engineer may share with his / her colleagues must not come with time and expenses of others.

Answer to question 3

Eager is an employee of Engco, but he/she undertakes work on weekends and during evenings from SmallBox Corp. Such part-time employment is known as "moonlight". In fact, it is ethical to work for more than one employer, but it requires determination and stamina. You are required by the Code of Ethics that while you are moonlighting, you must disclose the situation to the original employer. In addition to this, you should not compete with the original employer and also, moonlighting should not result in reduction of your efficiency during your usual work hours for the original employer. In the given case, Eager is competing with his/her employer, Engco, when eager accepted the work at an hourly rate of 50% of the rate that Engco would have charged SmallBox. "The extra money would be nice", is not a convincing reason. Eager is a professional engineer and he/she can therefore, insist on receiving an adequate pay for his/ her professional work from Engco. Also, Eager has not informed Engco about his/her part-time job, then Eager's conduct is obviously unethical.

Answer to question 4

You are a professional engineer and you are therefore duty bound, in accordance with the Code of Ethics, to act as a guardian to protect health and properties of the general public. You must, at any time, consider safety and welfare of the public as paramount. You did a good job when you sent an e-mail to Potion. In the given case, though the error is reduced to 3% from 20% of incorrectly labeling the bottles, but 3% error (to your opinion) should be enough to put health of the drug users in jeopardy. You as a professional engineer must take an immediate action. The best way to begin with, is to write letters to the presidents of both the firms and draw their attention to their unethical conducts. Since most of the firms are honest, they most probably may comply with your recommendations and guidelines, if not the problem falls back on you. The reason "a two - month delay is not acceptable" must not be ethically acceptable to you. You must immediately inform the Ministry of Health as well as other concerned authorities. If due to consumption of the drug a serious event occurs (death, other health injury) then, you would be probably investigated for possible unethical behavior, or even collusion with the firms. A professional engineer who discovers that his / her employer is not honest must immediately dissociate himself / herself from any action contrary to the Code of Ethics. You may even blow the whistle if the firm's authorities completely ignore your opinions as well as your instructions. The conduct of Elixir's president is absolutely unethical and due to his / her negligent conduct he most probably would be disciplined.

ASSOCIATION OF PROFESSIONAL ENGINEERS OF ONTARIO

PROFESSIONAL PRACTICE EXAMINARTION-April 17, 2004

PART "B" – Engineering Law and Professional Liability

This examination comes in two parts **(Part "A" and Part "B")**. <u>Both </u>parts must be completed in this sitting. You will be given a total of **180 minutes** to complete the examination.

Use the correct colour – coded Answer Book for each part, place in the correct envelope and **seal after completed.**

> *White Answer Book for Part A white question paper:*
> *Coloured Answer Book for Part B coloured question paper:*

This is a **"CLOSED BOOK"** examination. **No** aids are permitted other than the excerpts from the 1990 Ontario Regulation 941 covering sections 72 (*Professional Misconduct*) and 77 (*Code of Ethics*) supplied at the examination. Dictionaries are **not** permitted.

The marking of the questions will be based not only on academic content, but also on legibility and the ability to express yourself clearly and correctly in the English language. If you have any doubt about the meaning of a question, please state clearly how you have interpreted the question.

All **four** questions constitute a complete paper for Part "B". Each of the four questions is worth 25 marks.

Front Page

4.3
(MARKS)

(25)   1.   <u>Briefly</u> define and explain any <u>five</u> of the following :

      (i)   The limitation period applicable to contracts in Ontario after January 1, 2004.
      (ii)   Five examples of inappropriate conduct in the workplace (list only)
      (iii)   Contract frustration
      (iv)   Fraudulent misrepresentation
      (v)   Vicarious liability
      (vi)   Economic duress
      (vii)   The Statute of Frauds
      (viii)   Gratuitous promise

(25)   2.   A joint venture consisting of both engineering and contracting firms entered into a contract with an Ontario city to design and build an all-electronic toll highway "expressway" featuring both underground tunnel portion and surface portions of the highway. The contract also required the joint venture to design, install and implement an electronic tolling system to accommodate specified numbers of vehicles, all as specified in the request for proposal for the design and construction of the all-electronic expressway, as published by the city.

      The contract between the city and the joint venture provided that all-electronic highway expressway was to be fully operational by a specified date, failing which the joint venture contractor would be responsible to pay the city liquidated damages (based on lost toll revenues in accordance with the project's feasibility study and financial plan) of $300,000 for each day beyond the specified completion date until the expressway and its all-electronic tolling technology was finally installed and fully operational. The contract also included a provision limiting the contractor's liability for liquidated damages under the contract to maximum amount of $30,000,000.

With the city's approval the joint venture contractor then subcontracted, to a firm specializing in tolling technology, the obligations to design, install and implement the tolling technology system as required by the city's specifications. The subcontract contained a provision obligating the tolling technology subcontractor to be responsible to the joint venture contractor to provide a fully operational tolling system by same specified date and for the same $300,000 of the daily liquidated damages (subject to the same maximum amount of $30,000,000 in liquidated damages) as set out in the joint venture contract between the joint venture contractor and the city

Although the expressway was otherwise operational by the specified completion date, the tolling technology subcontractor experienced difficulties in completing the installation and implementation of the tolling technology in accordance with the requirements of the subcontract. In fact, the tolling technology subcontractor was 120 days late in successfully completing the design, installation and implementation of the tolling technology system as required by the subcontract (and the Contract).

Explain and discuss what claim the joint venture contractor could make against the tolling technology subcontractor in the circumstances. In answering, explain the approach taken by the Canadian courts with respect to contracts that limit liability and include a brief summary of the development of relevant case precedents.

(25)   3.   Ontario Industrial Laundry Inc. (OILI") owns several laundry plants in Ontario. OILI's operations include handling the laundry for various customers around the province. OILI decided to build a large new plant in Brampton to replace a number of smaller and aging OILI facilities.

OILI engaged an architectural firm, Clever and Really Useful Design Developments Inc. ("CRUDDI"), and entered into an architectural services agreement with it. Under the agreement, CRUDDI was to design the new plant and prepare plans and specifications necessary to build it. According to the agreement, CRUDDI was to design "the most modern and technically up – to – date laundry in Canada"

CRUDDI hired several consultants to provide the various services necessary for the project. Of these, Mechanical Engineering Systems and Services Inc. ("MESSI") was to design the air conditioning and handling system.

Although MESSI did not have a contract with OILI, it worked closely with a representative of OILI who specified that, as it was important to provide comfortable working temperatures in the plant, the air conditioning and handling system must be able to provide working temperatures in the range of 22° C and 25° C and a minimum of 18 air changes per hour.

OILI, on the basis of competitive tenders, awarded the contract for the construction of the new plant to Dominion Industries and Related Technologies Inc. ("DIRTI"). The contract price was $15,000,000. DIRTI completed the construction in accordance with the contract drawings and specifications.

Almost immediately after having commenced its

operation in the new plant, OILI experienced problems in the air conditioning and handling system. The temperature in the working areas was excessive, reaching 38° C in the summer months. In the compressor room, the temperature reached 50° C and caused malfunctions In addition, the circulation was poor and the air quality was offensive. The employees began suffering fatigue and other ailments and it became necessary for them to take frequent "heat breaks".

CRUDDI and MESSI tried several times to remedy the problems but they were unsuccessful. OILI retained Top Industrial Designs Inc. ("TIDI"), another mechanical engineering company, to conduct an independent investigation. TIDI determined that the air conditioning and handling system was under designed. The air conditioner's chilling unit had a capacity of only 230 tons, a larger unit having a capacity in the order of 600 tons should have been specified. In addition, the exhaust and intake vent on the roof were located too close to each other and caused exhausted air to re-enter the plant.

TIDI determined that the system would require $1.1 million in modifications In order to meet plant's specifications. It also indicated that, had the system been specified and constructed, as it ought to have been in the first place, construction costs incurred by OILI would have been $400,000 higher, that is, $15400,000.

What potential liabilities in tort law arise in this case? In your answer, explain what principles of tort law are relevant and how each applies to the case. Indicate a likely outcome to the matter.

(25)   4.   A mining contractor signed an option contract with a land owner which provided that if the mining contractor (the "optionee") performed a specified minimum amount of exploration services on the property of the owner (the "optionor") within a nine month period, then the optionee would be entitled to exercise its option to acquire certain mining claims from the optionor.

Before the expiry of this nine-month "option period", the optionee realized that it couldn't fulfill its obligation to expend the required minimum amount by the expiry date. The optionee notified the optionor of its problem prior to expiry of the option period and the optionor indicated that the option period would be extended. However, no written record of this extension was made, nor did the optionor receive anything from the optionee in return for the extension.

The optionee then proceeded to perform the services and to finally expend the specified minimum amount during the extension period. However, when the optionee attempted to exercise its option to acquire the mining claims the optionor took the position that, on the basis of the strict wording of the signed contract, the optionee had not met its contractual obligations. The optionor refused to grant the mining claims to the optionee.

Was the optionor entitled to deny the optionee's exercise of the option? Identify the contract law principles that apply, and explain the basis of such principles and how they apply, to the positions taken by the optionor and by the optionee.

4.4 Suggestive answers to Law, APRIL 2004

Answer to question 1

(i) Limitation period is a specified period of time during which a claim for the damages can be made. If a claim is made after expiry of the period, then the claim would not succeed. This is known as "statute barred".

(ii) The inappropriate conducts in the work place as have been prohibited by Ontario Human Rights are: discrimination based on race, place of origin, skin colour, ethnic origin, sex, handicap and age. The Code also outlaws sexual harassment.

(iii) An unexpected and highly rapid change of circumstances may cause the party to a contract not to be able to fulfill his/her contractual obligations. When such exceptional circumstances occur, then the contract is frustrated and is discharged by frustration. Occurrence of war is a good example. However, the concept of discharge by frustration cannot be applied to justify discharge of a contract by frustration due to changes of circumstances, which were beyond their contemplation.

(iv) A misrepresentation means a statement of false. There are two types of misrepresentations: innocent and fraudulent. Fraudulent misrepresentation is one which a party to a contract carelessly and on purpose misleads the second party to induce him to a contract. The deceived person can discharge the contract and he is also entitled to claim compensation for damages. He can sue for the damages.

(v) Vicarious means acting on behalf of somebody else. When an owner of a property sustains damages due to negligent conduct of an employee of a firm, then his/her employer would be vicariously liable for the damages.

(vi) Duress means intimidation. It is when threat or violence is used to induce a person to an unfair contract. When this threat or violence causes severe economic loss, then economic duress is involved. This type of contract is not legitimate and therefore it is not enforceable.

(vii) A contract may be verbal or written. A written contract may be formed in part through negotiation and in part by discussion. A written contract must include all the terms that were mutually agreed upon by the parties. The statute of frauds indicates that certain type of contract must be in writing to be enforceable. Such as:
- i-   contracts relating to interest in a land
- ii-   those agreements that cannot be completed within a period of one year
- iii-   guarantees of indebtedness

(viii) It is a promise without consideration. A promise or a contract without consideration cannot be enforced. An exception exists in the case of equitable estoppel .

Answer to question 2

This case is dealing with fundamental breach of contract. The contract between the joint venture and subcontractor indicates that there would be a daily-liquidated damage of $ 300,000. The contract also includes an exemption clause limiting liability of subcontractor to $ 30,000,000.

At one time Canadian courts followed English courts precedent in applying what was called the "fundamental breach doctrine". According to this doctrine, an exemption clause cannot be enforced in case of a fundamental breach. Eventually, this doctrine was overruled by English courts. Even though, the concept of this doctrine is not completely overruled by Canadian courts, but strong preferences are given to the exemption clauses in commercial cases. Canadian courts following English courts do not disregard the exemption clauses.

In the given case, subcontractor was 120 days late therefore according to this contract the joint venture should have received 120*300,000 = 36,000,000 dollars. But since subcontractor has limited his total liability to $ 30,000,000, the joint venture would be entitled to receive only $ 30,000,000.

Answer to question 3

The essential purpose of tort which mean damage and / or injury, is to compensate a party who has sustained damages / injury as a result of negligent conduct of another party. In order to succeed in tort claim, there is no need that the two parties must be in a contractual relationship. The plaintiff must prove that:
1-   the defendant owed the plaintiff a duty of care.
2-   the defendant breached that duty through his negligent conduct.
3-   The plaintiff has sustained damages as result of the defendant negligent conduct.

In the given case, the plaintiff would be OILI and the defendant would likely be MESSI. MESSI duty of care was to design an air conditioning and handling system to provide comfort to the Employees. MESSI knew through a representative of OILI that it was important to provide comfortable working temperature. All the required information and specifications about air conditioner's temperature were given to MESSI. MESSI knew or ought to have known that full reliance has been kept on it. MESSI should have foreseen the damages that may arise as a result of its negligent conduct. MESSI breached that duty of care since the temperature in the working area was excessive (38°C instead of 22°C to 25°C).

Investigations by TIDI confirmed that the air conditioning and handling system were under designed. That means, MESSI breached his duty of care, which MESSI owed OILI. Claim for tort liability can arise when damage manifests itself, excessive temperature resulting in frequent "heat break" by the employees. MESSI may also be liable for putting the employees health in danger. OILI most probably might have also suffered economic loss.

Answer to question 4

This is the case, which involves "equitable estoppel", which states that when a party to a contract makes a gratuitous promise, a promise without consideration, and the second party to the contract reasonably relies on it, then it would be unfair and inequitable to allow the first party to revert to his contractual terms. Court always enforces only those terms and conditions that are clearly and unambiguously written in a contract and a contract without consideration cannot be enforced. If a term has verbally been agreed upon, but has not been entered into the contract, then this term is not enforceable. But in case of a gratuitous promise, since it would be unfair and inequitable to let the first party return to his strict contractual term, the concept of equitable estoppel is well applicable.

In the given case, the "optionor" made a gratuitous promise when he indicated to the "optionee" that the option would be extended. The "optionee" reasonably relied on it. Therefore, the"optionor" must be estopped. No, the "optionor" was not entitled to deny the optionee's exercise of the option.

CHAPTER 5

PPE, Part "A", Part "B", August 2004

Abstract. Attempts have been made to provide the readers with full clarity of the language adopted to answer the questions of Part "A" - Professional Practice and Ethics, and Part "B" – Engineering Law and Professional Liability of August 2004.

ASSOCIATION OF PROFESSIONAL ENGINEERS OF ONTARIO

PROFESSIONAL PRACTICE EXAMINARTION- August 14, 2004

PART "A" – Professional Practice and Ethics

You will be given a total of **90 minutes** to complete this examination.

Use the correct colour – coded Answer Book for each part, place in the correct envelope and **seal after completed.**

> *White Answer Book for Part A white question paper:*
> *Coloured Answer Book for Part B coloured question paper:*

This is a "**CLOSED BOOK**" examination. **No** aids are permitted other than the excerpts from the 1990 Ontario Regulation 941 covering sections 72 (Professional Misconduct) and 77 (Code of Ethics) supplied at the examination. Dictionaries are **not** permitted.

The marking of the questions will be based not only on academic content, but also on legibility and the ability to express yourself clearly and correctly in English language. If you have any doubt about the meaning of a question, please state clearly how you have interpreted the question.

All **four** questions constitute a complete paper for Part "A". Each of the four questions is worth 25 marks.

Where a question asks if a certain action by an engineer was ethical or not, a simple "yes" or "no" answer is not sufficient. You should identify the applicable clauses in Regulation 941 and comment on the action in each situation.

Any similarity in the question to actual person or circumstances is coincidental.

5.1
Question 1

(10) **(a)** Professional engineering in Ontario is described as a "self-regulating profession". What doe this term mean? In your answer, briefly describe three different features in the way professional engineering is regulated in Ontario that are consistent with this term.

(5) **(b)** In order to be designated as a "Consulting Engineer" one must meet a number of requirements. Briefly list three of them. What additional privileges or rights are granted by this designation?

(5) **©** Describe the roles performed by PEO's Complaints Committee and its Discipline Committee.

(5) **(d)** Is a civil engineer allowed to perform services that are normally within the scope of mechanical engineering? Explain.

(25) **Question 2**

N. Trepreneur, a senior professional engineer, established a small firm, Trepreneur Engineering, to provide professional engineering services to the public. The firm became busy very quickly and within a few months, hired J.R. Green a bright, recent university graduate with an engineering degree, to assist with the work. N.Trepreneur strongly believed in mentoring and hoped that in several years, after obtaining the necessary experience requirements and becoming a P. Eng., Green would assume increasing managerial responsibility and possibly an ownership interest in the firm.

About a year after Green joined the firm, Trepreneur Engineering was asked by one of its clients to provide a formal report that included an engineering opinion. Green performed the work on that matter and prepared a draft of the report. Before having a chance to review Green's work, N.Trepreneur received an urgent request from another client that required N.Trepreneur to leave on a lengthy business trip. On the way out of the office, N.Trepreneur stopped at Green's desk and said, "Sorry, but I'll be out of the country and tied up completely for the next three weeks, so I won't be able to review that report. I know that it's due tomorrow, so go ahead and sign it under your own name and send it to the client so we meet the deadline". N.Trepreneur was confident that that would be alright, since Green had always produced outstanding work in the past. Green proceeded to complete the report, signed it "J.R. Green, Eng., Trepreneur engineering" and sent it to the client.

Discuss the conduct of N.Trepreneur and J.R. Green. What, if anything, should they be concerned about? Could N.Trepreneur and / or J.R. Green be disciplined by the Discipline Committee of the PEO? Is there anything about N.Trepreneur's conduct relative to the Code of Ethics that is commendable? <u>In your answer, please assume that Green's report would have no impact on public safety or welfare.</u>

Any similarity in the question to actual person or circumstances is coincidental.

(25)   Question 3

You are a professional engineer and have been hired recently as the chief operations and maintenance engineer of a paper mill near a remote village in northern Ontario. The mill is the largest industry in the area and employs (directly or indirectly) most of the workers in the region.

Upon starting your new job, you review the facilities at the mill and its operation and maintenance procedures. You discover that your predecessor (also professional engineer) had been operating the mill for several years with inadequate environmental equipment. The mill has been discharging hazardous substances into a nearby river contrary to legal limits. You also learn that government authorities are not aware of the illegal discharges.

You discuss the situation with the company's vice president in charge of Canadian operations. You report to the vice president that a number of environmental measures would be necessary in order to stop the illegal discharge. In your estimation, the measures would require a substantial capital investment in the plant. The vice president inform you that the mill has been earning very small profits in the last decade and that the capital expenditure could not be justified at this time. According to the vice president, the company's head office would likely close the mill rather than spend the money. You are told that some of the environmental measures you are proposing could be implemented gradually as the mill's financial performance improves.

        (a)   Discuss your obligations with respect to the company.

        (b)   Discuss your obligations with respect to the public. What is the public interest in this case? How is the public interest impacted by your actions? Comment on the vice president's promise to gradually implement the environmental measures. In addition to the measures you propose, what other steps should be taken with respect to the public interest?

        ©   Discuss your obligations with respect to your predecessor.

(25)   **Question 4**

D.U. Plicitous, P. Eng., is employed by a municipality in Ontario as head of the municipality's procurement department . Plicitous's responsibilities include establishing procurement policies and procedures for the municipality as well as participating in the bid selection and contracting process.

The municipality is currently considering hiring a company to design and build a wastewater treatment facility. A key feature of the facility would be the ability to stabilize and dewater the municipal sludge to transform it into a useful fertilizer and soil conditioner.

The municipality's staff has prepared a draft Request for Tenders for the project. Before it is issued to prospective bidders, it is reviewed by Plicitous. Plicitous is generally satisfied with the draft and makes only a few revisions, including revisions to the scoring formula used to select the winning bidder. The current formula awards points based on price and compliance with various technical requirements in the Request for Tenders. According to Plicitous's revisions, up to 10 points could be awarded based on the amount of experience the bidder has in designing and building such projects, and local bidders would receive 10 points automatically.

Plicitous chairs a committee charged with evaluating scoring and selecting the winning bidder. Of the bids received, ABC and XYZ received the most points from the committee as described in the table below:

	Possible Points	ABC's Score	XYZ's Score
Technical	40 points	35 points	35 points
Price	40 points	28 points	40 points
Experience	10 points	10 points	3 points
Local Bidder	10 points	10 points	0 points
Total	100 points	83 points	78 points

Although ABC and XYZ have similar experiences, XYZ's score was reduced to 3 points for experience because, according to statements made by Plicitous at the committee, XYZ's engineers had produced a poor design on one of its previous projects. In addition, ABC was the only local bidder. The committee informed ABC that it had won the job.

Later that evening, Plicitous was treated to a celebration dinner at an expensive restaurant by A. Complice. Plicitous is married to Complice, the president of ABC.

Comment on Plicitous's conduct.

Any similarity in the question to actual person or circumstances is coincidental.

5.2 Suggestive answers to Ethics, AUGUST 2004

Answer to question 1

(a)   Association of Professional Engineers regulates profession of engineering. Majority of members of the Association's Council are elected by and from the members of the Association. Association's Regulations and by laws are approved by the Council. This Council also directs the staffs who administer the Professional Engineers Act and Regulations. The disciplinary committee formed by licensed engineer are appointed by the same Council to disciplines Professional Engineers. Since engineers themselves govern the engineering profession, it is, therefore called "self – regulating".

(b)   In order to be designated as a "Consulting Engineer", the member must
  1- have two years experience in private practice
  2- pass examinations assigned by the Association Council or be exempted
  from it.
  3- must have five (5) years experience since becoming a member.
Since the applicant is engaged in private practice and therefore offering his/her services to the public, he/she is required to have a Certificate of Authorization or must be associated with a corporation that is holder of the certificate.

©   The Complaints Committee evaluates the complaint. This committee is composed of members of the Association's governing council and other licensed members. Discipline Committee conducts a formal hearing that renders judgment. This committee is composed of members of governing council and people who have not participated in the previous steps.

(d)   If a civil engineer is engaged in performing works, which is within the scope of mechanical engineering, then surely he/she would be charged with negligence and incompetence, which may be basis for discipline under every provincial Act .

Answer to question 2

When a person who is not licensed by the PEO infringes the Professional Engineers act and practices engineering without licence or uses an engineering seal and the title "Professional Engineer" or any title that may make people believe that he/she is really a Professional Engineer, has violated the Professional Engineering Act. This person will be prosecuted, in accordance with the Act, in the court. If convicted, he/she may have to pay severe fines. In the given case, J.R.Green, signs the report "J.R.Green, Eng., which may make the client believe that he is a professional engineer.
N. trepreneur is a professional engineer, he is therefore duty bound to safeguard welfare of the public and must always consider safety of the general public as paramount. When N.Trepreneur tells Green "Go ahead and sign it", it is as good as approving the report that was not based on his thorough knowledge. Such an action may cause Trepreneur to be guilty of professional misconduct. Professional misconduct is the main cause for being discipline by Discipline Committee.

Answer to question 3

(a)   When you are employed by a firm, you are duty bound to utilize your ability and knowledge to obtain your employer's legitimate goal. You must keep his business confidential. In a rare case your employer may ask you to act contrary to the welfare of the society. Such an action is unethical and may be illegal. As a professional engineer, you must not only refuse to perform such unethical activities, but you may have to defend or explain your refusal.

(b)   You did a good job to discuss the situation with the company's vice president. However, when he failed to respond adequately, the responsibility fell back on you. You are a professional engineer and you are duty bound by the Code of Ethics to protect the environment. You must consider safety and welfare of the general public as paramount. You should insist that unsafe performances are not acceptable. You must inform the mill authorities that "the mill has been earning very small profits and the environmental measures could be implemented gradually" are not acceptable. You, as a professional engineer, are the ultimate authority and must insist on protecting the environment as well as the general public, especially when hazards may threaten health of the people and / or cause other damages. According to the Code of Ethics, failure to protect the public and environment may cause disciplinary action.

(c)   Your predecessor was a professional engineer and had been operating the mill for several years with inadequate environmental equipment. As a result putting health and safety of the public and possibly other specious in danger. He has failed to comply with the Code of Ethics and therefore must be disciplined.

Answer to question 4

Accepting a gift could be an indication of a conflict of interest. Plicitous created a serious conflict of interest when he accepted the invitation to an expensive dinner. It is strictly prohibited to exchange a gift in any kind of tendering process. Such conduct may constitute bribery. Plicitous chairs a committee to evaluate and select the winning bidder, therefore, he has put himself in a very awkward position. Another conflict of interest that Plicitous has created for himself is that he is married to the president of ABC. Plicitous should have revealed the fact that he was married to the president of ABC and consequently, Plicitous should have withdrawn himself / herself from chairing the committee charged with evaluating and selecting the winning bidder.

ASSOCIATION OF PROFESSIONAL ENGINEERS OF ONTARIO

PROFESSIONAL PRACTICE EXAMINARTION-August 14, 2004

PART "B" – Engineering Law and Professional Liability

You will be given a total of **90 minutes** to complete this examination.

Use the correct colour – coded Answer Book for each part, place in the correct envelope and seal after completed.

>   *White Answer Book* for Part A white question paper.
>   *Coloured Answer Book* for Part B coloured question paper.

This is a "**CLOSED BOOK**" examination. **No** aids are permitted other than the excerpts from the 1990 Ontario Regulation 941 covering sections 72 (*Professional Misconduct)* and 77 (*Code of Ethics*) supplied at the examination. Dictionaries are not permitted.

The marking of the questions will be based not only on academic content, but also on legibility and the ability to express yourself clearly and correctly in the English language. If you have any doubt about the meaning of a question, please state clearly how you have interpreted the question.

All **four** questions constitute a complete paper for Part "B". Each of the four questions is worth 25 marks.

Front Page

5.3
(MARKS)

(25)　1.　<u>Briefly</u> define and explain any <u>five</u> of the following:

(i)　　Statutory holdback
(ii)　　Five examples of inappropriate conduct in the workplace (list only)
(iii)　Liquidated damages
(iv)　Rule of contra proferentem
(v)　　Secret commission
(vi)　Innocent misrepresentation
(vii)　Exemption clause
(viii)　Director's standard of care

(25)　2.　A long established manufacturing company, Acme Ltd., contemplating the possibility of a sale of some of its properties, retained an environmental consulting firm, E Inc. to prepare an environmental compliance audit.

The Vice-President of E Inc. responsible for the performance of the environmental compliance audit, a professional engineer, turned the matter over to one of the department's engineering employees. The engineering employee in question to whom the matter was referred had only recently qualified as a professional engineer. However, on the basis of previous assignments, the Vice-President had been very impressed by the young engineer's abilities. The Vice-President was also aware that the department's extremely busy schedule would likely limit the amount of time the Vice-President could spend on the environmental compliance audit and, accordingly, selected the younger employee engineer in the hope that the junior engineer's involvement, particularly in view of the junior engineer's impressive performance on previous matters, would decrease the Vice-President's supervisory time in connection with the audit.

The employee engineer carried out an environmental compliance audit with respect to each of the properties identified and E Inc. submitted its reports on each property. Included at the outset of each report was the following qualifying statement:

"This report was prepared by E Inc. for the account of Acme Ltd. The material in it reflects E Inc.'s best judgment in light of the information available to it at the time of preparation. Any use which a third party makes of this report, or any reliance on decisions to be made based on it, are the responsibility of such third parties. E. Inc. accepts no responsibility for damages, if any, suffered by any third party as a result of decisions made or actions based on this report."

1 of 4 pages　　　August 14, 2004.　Part B – PPE

Some time later, Acme Ltd. sold tow of its properties to Acquisitions Inc. In negotiating the sale with Acquisitions Inc., E. Inc.'s reports were shown to Acquisitions Inc., but Acquisitions Inc. had no dealing with E. Inc.　E Inc. had no knowledge of the sale to Acquisitions Inc. until approximately four years later when Acquisitions Inc. commenced a lawsuit against E. Inc. Acquisitions Inc. claimed it had commenced the lawsuit in tort against E. Inc. because it had encountered hazardous substances on one of the properties and had subsequently obtained the opinion of another environmental consulting firm who confirmed that the report in question by E Inc. contained negligent misstatements which, in the opinion of the second consulting firm, had resulted from E. Inc.'s representatives having spent too little time investigating the property for hazardous substances. Acquisitions Inc. claimed in its lawsuit that E. Inc. was aware that the report might be shown to prospective purchasers and, accordingly, E. Inc. should be responsible for damages arising as a result of reliance by Acquisitions Inc. on the negligent misstatements in E. Inc.'s report.

What potential liabilities in tort law arise in this case? In your answer, explain what principles of tort law are relevant and how each applies to the case Indicate a likely outcome to the matter. In your answer indicate if your conclusion would differ if the reports by E. Inc. had not contained the qualifying statement identified above and, if your conclusion would differ, explain why.

3.   Equipment Inc. ("Equipment") is a company engaged in the business of supplying heavy equipment used in the oil exploration and drilling industry.

Equipment had become aware that Oil Company Ltd. ("Oilco") required a contractor to design, manufacture, supply and install specialized gear boxes. The gear boxes would be used to drive a number of bucket wheel conveyor belts that transported sand at Oilco's oil extraction tar sands plant in Alberta. Equipment decided to tender on the Oilco contract.

In order to tender on the contract, Equipment set out to purchase the gear boxes. Equipment was contacted by a representative of Manufacturing Ltd., ("Manufacturing")a company which manufactured similar equipment. After visiting Oilco's site and examining the conveyors, the representative of Manufacturing became familiar with the requirements of the gear boxes. Manufacturing represented to Equipment that Manufacturing would be able to design and manufacture the specialized gear boxes and that the gear boxes would be suitable for the purpose intended. On the basis of these representations, Manufacturing and Equipment entered into a contract. Manufacturing agreed that if Equipment was successful in its tender to Oilco, Manufacturing would provide the equipment for a price of $600,000. The contract also contained a provision limiting Manufacturing's total liability to $600,000 for any loss, damage or injury resulting from Manufacturing

2 of 4 pages   August 14, 2004.   Part B – PPE

performance of its services under the contract.

Based on the information provided by the Manufacturing representative, Equipment prepared and submitted its tender to Oilco. Oilco accepted the tender and entered into a contract with Equipment for the gear boxes.

The gear boxes were installed at Oilco's site by employees of Equipment according to Manufacturing's installation procedures. Shortly after the gear boxes were put into service, main gears inside them failed. As a result of this failure, the conveyors were damaged and it was impossible for Oilco to operate its conveyors. Manufacturing made several unsuccessful attempts to correct the gear boxes.

In order to meet its obligations under the Oilco contract, Equipment hired another supplier to correct the defects in the gear boxes. For an additional $800,000 Equipment was able to correct the problem by replacing the gear boxes with gear boxes manufactured by another company and by repairing the damage to the conveyors. The total amount which had been paid by Equipment to Manufacturing prior to discovering the defects was $450,000.

Explain and discuss what claim Equipment can make against Manufacturing in the circumstances. Would Equipment be successful in its claim? Why? In answering, please include a summary of the development of relevant case precedents.

(25) 4.   A newly formed energy company ("NEWCO") decided to investigate the possibility of developing a liquefaction process to convert coal deposits into oil.

NEWCO entered into a contract with a large engineering firm pursuant to which the engineering firm was to carry out a feasibility study to determine, over a period of eight months and by a specified date, the feasibility of the proposed liquefaction process. The contract between NEWCO and the engineering firm expressly provided that should the feasibility study be completed by the "deadline" date specified and should the results of the study indicate that the liquefaction process proposed by the engineering firm would meet the specified quality and volumes of liquefied oil output, then the engineering firm would be authorized to carry out further work to develop the liquefaction process to operate on a commercial basis, all on terms and conditions clearly set out in the contract between NEWCO and engineering firm.

The engineering firm undertook the feasibility study and, although the results of the feasibility study appeared promising and in compliance with the parameters specified in the contract with NEWCO, the engineering firm found that it would be unable to complete the feasibility study by the date specified.

The president of the engineering firm explained to the president of the NEWCO that the engineering firm would not be able to fulfill all aspects of the feasibility study as required by the specified date. The president of the engineering firm emphasized that whereas the engineering firm would likely be two weeks late in completing its feasibility study obligations, the results of the feasibility study indicated that the liquefaction process would very likely meet NEWCO's requirements for commercial production as specified.

The president of NEWCO indicated to the president of the engineering firm, verbally, that the time for completion of the feasibility study would be extended.

The engineering firm completed the feasibility within two weeks after the date specified in the contract.

Subsequently, NEWCO took the position that the engineering firm had not completed the feasibility study in time and, accordingly, that NEWCO was not obligated under the wording of the contract to authorize the engineering firm to carry out further work to develop the liquefaction process on a commercial basis. Instead, NEWCO issued a request for proposals from several firms for the development of the liquefaction process to operate on a commercial basis. NEWCO selected another firm that was prepared to undertake the development of the process for a fee substantially lower than the fee that was to have been paid to the original engineering firm had it completed the feasibility study by the date specified in the contract.

Was NEWCO entitled to deny the engineering firm's right to develop the liquefaction process to operate on a commercial basis? Identify the contract law principles that apply, and explain the basis of such principles and how they may apply to the position taken by NEWCO and by the engineering firm.

5.4   Suggestive answers to Law, AUGUST 2004

Answer to question 1

(i)   This is a term, which is commonly used in construction contract. It is a percentage of money held back by the owner from the contractor. In Ontario it is 10% of the total price. There are two types of holdback, 1-basic holdback or substantial performance, 2-finishing holdback. The owner is obligated to retain the holdback until he receives a notice that claims for lien related to holdback is satisfied, expired or discharged. If the owner releases the holdback money too soon, then he may have to pay extra fund to satisfy the lien claimants.

(ii)   The inappropriate conducts in the work place as have been prohibited by Ontario Human Rights are: discrimination based on race, place of origin, skin colour, ethnic origin, sex, handicap and age. The Code also outlaws sexual harassment.

(iii)   Liquidated damages or pre- estimated damage is a clause in a contract. This clause pre- estimates damage, which is likely to occur due to non-performance of a contract. It is also the clause by which a contractor limits his liability for tort damages.

(iv)   It is a rule that indicates the importance of a clear and unambiguous language in preparing a contract. This rule emphasizes that a contract must be free from ambiguity. If a contract contains ambiguity then it will be interpreted against the party who poorly drafted it.

(v)   It is money, or something of value that has been offered through a secret agreement to show a favour or disfavour to a third party. Secret commission is an offence under Criminal Code of Canada. If a Professional Engineer is involved in secret commission, then he/she will be charged under Regulation 941/90.

(vi)   An innocent misrepresentation is a lie, a statement of false made by a person who truly believes that his/her statement is true. The deceived party can repudiate the contract and is entitled for compensation.

(vii)   It is a provision by which the parties to a contract limit their liabilities.

(viii)   Every director or officer of a corporation must use his / her power
     a- with honesty and in good faith to protect interest of the corporation
     b- with care, diligence and skill that a prudent person would exercise in that circumstances.

Answer to question 2

Tort means damage, injury and its main purpose is to compensate a person who has sustained damages as a result of negligent conduct of another person. Claim for tort may arise, even though the two parties are not in contractual relationship. In order for a claim for tort to be successful the plaintiff must prove that:

a-   the defendant owed the plaintiff a duty of care.

b-   the defendant breached that duty by his negligent conduct

c-   the defendant conduct caused damages to the plaintiff.

In the given case, the plaintiff is Acquisition and the defendant would likely be the young professional engineer and ultimately his firm, the E. Inc. As a Professional Engineer, he knew or ought to have known that Acme Ltd has put its full reliance on his professional report. Professional Engineer is one who must consider safety and welfare of general public as paramount. He must always foresee the damages that may arise as a result of his/her negligent conduct. He owed a duty to prepare an environmental compliance audit. The investigation by another environmental consulting firm indicated that the Professional Engineer had spent too little time He neglected his duty by such conduct. As a Professional Engineer, he is duty bound to make a thorough and accurate investigation. Claim for tort can arise when the damage manifests itself, existence of hazardous substances.

Courts always honour the exemption clause disclaiming responsibility for any damages suffered by the third party. Since the report prepared by the Professional Engineer contains the exemption clause, which is clearly and unambiguously disclaiming any responsibility for damages, if any, suffered by any third party therefore, Acquisition is not entitled to any claim. But if the clause disclaiming responsibility to the third party were not included, then the Professional Engineer would be liable to compensate for the damages. Since he is an employee of E. Inc., his employer would be vicariously liable to the damages.

Answer to question 3

This case is dealing with fundamental breach of contract. The contract between Manufacturing and Equipment indicates that there is an exemption clause limiting total liability of Manufacturing to $ 600,000.

At one time Canadian courts followed English courts precedent in applying what was called the "fundamental breach doctrine". According to this doctrine, an exemption clause cannot be enforced in case of a fundamental breach. Eventually, this doctrine was overruled by English courts. Even though, the concept of this doctrine is not completely overruled by Canadian courts, but strong preferences are given to the exemption clauses in commercial cases. Canadian courts following English courts do not disregard the exemption clauses.

In the given case, Equipment has spent a total of 800,000 + 450,000 = 1,250,000 dollars for the job that should have been completed for $ 600,000. But since Manufacturing has limited its total liability to $ 600,000, and court will apply the clear wording of the exemption clause, Equipment would be entitled to receive only $ 600,000.

Answer to question 4

This is the case, which involves "equitable estoppel", which states that when a party to a contract makes a gratuitous promise, a promise without consideration, and the second party to the contract reasonably relies on it, then it would be unfair and inequitable to allow the first party to revert to his contractual terms. Court always enforces only those terms and conditions that are clearly and unambiguously written in a contract and a contract without consideration cannot be enforced. If a term has verbally been agreed upon, but has not been entered into the contract, then this term is not enforceable. But in case of a gratuitous promise, since it would be unfair and inequitable to let the first party return to his strict contractual term, the concept of equitable estoppel is well applicable.

In the given case, the president of NEWCO made a gratuitous promise when he, verbally, indicated to the president of the engineering firm that the time for completion of the feasibility study would be extended. The president of the engineering firm reasonably relied on it. Therefore, NEWCO must be estopped. No, NEWCO was not entitled to deny the engineering firm's right to develop the liquefaction process to operate on a commercial basis.

CHAPTER 6

PPE, Part "A", Part "B", December 2004

Abstract. This chapter provides you with the suggestive answers to the question papers of the both parts, Part "A"- Professional Practice and Ethics, and Part "B" – Engineering Law and Professional Liability of December 2004.

ASSOCIATION OF PROFESSIONAL ENGINEERS OF ONTARIO

PROFESSIONAL PRACTICE EXAMINARTION- December 11, 2004

PART "A" – Professional Practice and Ethics

This examination comes in two parts **(Part "A" and Part "B")**. Both parts must be completed in this sitting. You will be given a total of **180 minutes** to complete the examination.

Use the correct colour – coded Answer Book for each part, place in the correct envelope and **seal after completed.**

> *White Answer Book for Part A white question paper:*
> *Coloured Answer Book for Part B coloured question paper:*

This is a "**CLOSED BOOK**" examination. **No** aids are permitted other than the excerpts from the 1990 Ontario Regulation 941 covering sections 72 (*Professional Misconduct*) and 77 (*Code of Ethics*) supplied at the examination. Dictionaries are **not** permitted.

The marking of the questions will be based not only on academic content, but also on legibility and the ability to express yourself clearly and correctly in the English language. If you have any doubt about the meaning of a question, please state clearly how you have interpreted the question.

All **four** questions constitute a complete paper for Part "A". Each of the four questions is worth 25 marks.

Where a question asks if a certain action by an engineer was ethical or not, a simple "yes" or "no" answer is not sufficient. You should identify the applicable clauses in Regulation 941 and comment on the action in each situation.

Part A – PPE, December 11, 2004   Page 1 of 4

6.1
Question 1

(5)   (a)   Is there any difference between being a member of PEO and holding a licence to practice professional engineering in Ontario? Explain.

(5)   (b)   What are the consequences, if any, to a professional engineer who does not keep his or her licence permanently displayed in his or her place of business?

(5)   ©   What is the purpose of the engineer's seal and when should it be used? What two elements are required to accompany the seal?

(5)   (d)   Are there any restrictions on how professional engineering services may be advertised? Explain.

(5)   (e)   Does merely being designated as a "Consulting Engineer" allow a professional engineer to offer professional engineering services to the public? Explain.

(25)   Question 2

FastTrack Company is a design-builder and has been hired by a municipality in Ontario to engineer, procure and construct an automated light rail transit system (the "LRT").

You are employed as a professional engineer by FastTrack and are on the design team that is designing the LRT. The LRT's design includes a new, sophisticated computer-controlled system for the operation of its trains.

One night after work, you get together for dinner with an engineering colleague you met years ago, while at university. Your colleague also works at FastTrack. After a few drinks, your colleague commented that FastTrack was rushing ahead too fast with the LRT project under the pressure of a tight competition schedule. Your colleague also stated that FastTrack is not conducting sufficient testing of the LRT and is concerned that flaws in the system may go undetected until the LRT goes into operation.

The next day, you approach your direct supervisor, O. Verseer, P. Eng., to discuss these concerns. Your supervisor advises that your friend although very competent, is an alarmist. O. Verseer adds that with too much testing the project could be delayed unnecessarily causing it to be cancelled. Its cancellation would not only result in the loss of many jobs, it would also deprive the public of the benefits of a much needed transit system. Your supervisor reasons that all technological advanced entails some degree of uncertainty and risk and that insistence upon absolute safety would impede engineering progress.

In view of the above stated facts, what do you see as your ethical obligations as a professional engineer? Comment on O. Verseer's conduct. What do you think about your colleague's actions?

Question 3

WorldEng, a large engineering firm, was hired to prepare the design for a chemical production plant for MegaChem. In addition to preparing the plant design, WorldEng's duties included providing inspection services during the construction stage of the project. The project was completed successfully.

You are a P. Eng. and have been employed on a full – time basis by WorldEng for several years. You work in the Process Division and are involved on several process design projects. You were an important member of the design team that prepared the design for MegaChem's plant. In addition to working for WorldEng, you supplement your income by occasionally undertaking work on weekends and during evenings for EngInc, another engineering company. A colleague of yours, who is a P. Eng. at EngInc., assigns you such work and assumes responsibility for it.

A few years after the plant was completed, MegaChem decided to restructure its operations and sell the plant. BuyerCo has agreed to buy the plant, but before it does so, BuyerCo wants to satisfy itself (and its bank) that the plant was built to proper standards and is in good physical condition. BuyerCo hires EngInc. to inspect the physical plant and to review relevant documents (including the original plans and specifications, "as-built" drawings, and operations and maintenance logs). EngInc is very busy on several projects and asks you to assist with the plant inspection and document review.

(10) (a) comment on the appropriateness of your employment arrangements.

(10) (b) Assuming that your employment arrangements have not changed since the plant was designed and constructed, how do you respond to EngInc's request for assistance?

(5) © Is a P. Eng. licence sufficient to permit you to provide services to EngInc.? Explain.

(25)   **Question 4**

You are a professional engineer employed by FirstConcept Ltd., consulting engineering company.

One of the firm's clients, ShopCo Development Ltd. ("ShopCo"), has hired your firm to provide the engineering design for a new elevated walkway for one of ShoCo's major shopping malls. You have been assigned the responsibility of preparing the design for ShopCo.

You develop the design and meet with R. Epp, a representative   of ShopCo, to discuss your design. R. Epp is not a professional engineer, but disagrees with your design. The representative gives you some suggestions on how to simplify your design. You listen politely, but realize that the design would be compromised if the suggestions were incorporated.

Because of the disagreement, R. Epp fires FirstConcept and demands that you turn the design drawings over to A. L. Ternate, a professional engineer employed by SecondConcept Ltd., another engineering firm.

R. Epp had worked with A. L. Ternate on a previous project, and the two of them got along quite well. A. L. Ternate has agreed to complete the design of the new elevated walkway as R. Epp wishes.

You refuse to turn over the design drawings, even when R. Epp offers to pay for all of FirstConcept's services to date.

Do you have any obligation to turn over the drawings? Do you have any other responsibilities? Comment on A. L. Ternate's agreement with R. Epp.

6.2 Suggestive answers to Ethics, DECEMBER 2004

Answer to question 1

(a)   A licensee who is permitted by the Office of Professional Engineers Ontario to practice professional engineering in Ontario is also a member of Association of Professional Engineers Ontario.

(b)   A professional engineer must endeavor to keep his / her licence, temporary licence, limited licence, or certificate of authorizations, as the case may be, permanently displayed in his / her place of work.

©   According to professional engineering Act, every professional engineer must have a seal. This seal denotes that he / she is licensed. Only final documentation in the field of engineering must be signed and sealed. This seal, which has legal significance, indicates that the documentation has been prepared by a competent person and it identifies the responsible person. This seal must not be used for the documentations that neither have been prepared by the engineer nor been supervised by him/her.

(d) According to regulations made under the authority of the provincial Act, advertising is permitted in Ontario, but it should be done in a professional manner. It should be factual, it should not exaggerate. The advertisement should not criticize other person(s) in any way. Use of engineer's seal and seal of the Association is strictly prohibited.

(e) Since a "Consulting Engineer" is engaged in private practice and therefore offers his/her services to the public, the consulting engineer must be holder of "Certificate of Authorization" or he/she must be associated with a partnership or corporation that is a holder of the certificate. It is absolutely advisable for him to have professional liability insurance.

Answer to question 2

You are a professional engineer and therefore, pursuant to the Code of Ethics of your Association you are bound to consider safety and welfare of the general public as paramount. In the given case;
1.   your colleague commented on FastTrack and the LRT project after he had a few drinks
2.   You had not met your colleague for a long time
3.   He has been working at FastTrach (duration not mentioned)
4.   O. Verseer, your supervisor, advises you that your friend is an alarmist.
Consequently, several possibilities exist:

Case-a: Let us assume for a while that your friend was completely influenced by alcohol and he was not aware of what he was talking about. In addition to this, let us assume that your supervisor knew him much better than what you did (there has not been any meeting between you two for a long time) and your friend was indeed an alarmist, therefore there is no reason for you as professional engineer to rely on your friend's comments.

Case b- Let us assume that your friend was not influenced by alcohol and he was fully aware of what he was talking about. Let us also assume that your supervisor called your friend an alarmist because he wanted to avoid further testing, which surely results in spending time and money. In this case it is your sole duty to insist on conducting more tests until you are sure that the project is flaw-free and safety of the general public is fully protected.

It is advisable to meet your colleague friend immediately and be sure that he has not consumed any alcoholic beverages. But, if you are uncertain about the facts you must take action regardless of what your supervisor has told you. Under the Code of Ethics your supervisor has an obligation to respect your opinion.

Answer to question 3

(a) You are an employee of WorldEng, but you undertake work on weekends and during evenings for EngInc. Such part-time employment is known as "moonlight". In fact, it is ethical to work for more than one employer, but it requires determination and stamina. You are required by Code of Ethics that while you are moonlighting you must disclose the situation to the original employer. In addition to this, you should not compete with the original employer and also, moonlighting should not result in reduction of your efficiency during your usual work hours for the original employer. Since you have not informed you original employer about your part-time job, then your conduct is obviously unethical. Such disclosure enables your original employer to judge if your are competing with his work.

(b) This is a special type of conflict of interest. You are a full time employee of WoldEng. You were an important member of the design team that prepared the design for MegaChem. It will be judged as though you are doing a favour for yourself, if you assist with the plant inspection and document review. Your decisions may be based on impartiality, but it will not be observed to be fair by the general public. Once the truth is revealed, you will be exposed to conflict of interest.

Pursuant to the Code Ethics, a professional engineer must "avoid or disclose a conflict of interest that might influence the practitioner's action or judgment". So, in order to avoid this type of conflict of interest, you better inform EngInc of the fact that you are a full time employee of WorldEng and you participated in preparation of the design for MegaChem's plant. You better ask EngInc to send another person who does not have a conflict of interest. Failing to disclose a conflict of interest in awarding a major contract is a breach of the Code of Ethics.

© Since a colleague of yours, who is a P. Eng. at EngInc, assigns you the work and assumes responsibility for it, therefore it would be sufficient for you to provide services to EngInc, while you are moonlighting.

Answer to question 4

In the original documentation it should have been clearly mentioned, whether you or ShopCo owns the copyright for the drawing. But the case indicates that R. Epp is welling to pay for all FirstConcept's services, which clears the matter that ShopCo does not own the copyright of the drawing. Consequently, you do not have any obligation to return the drawing. You, as a professional engineer, are duty bound to safeguard the general public. You have listened politely, but have realized that the design would be compromised if the suggestions were incorporated. Under the Code of Ethics R. Epp has an obligation to respect your professional opinions. It is advisable to keep a client happy and satisfied with your work, provided that your client's suggestions and demands are reasonable and under no circumstances endanger the safety of the general public, because you must always consider safety of the general public as paramount.

There is nothing to be surprised that R. Epp has fired FirstConcept and has retained SecondConcept. It is a common practice by unsatisfied clients. A.L. Ternate is also a professional engineer. He has a similar duty towards the general public as you have. I do not see anything wrong with A.L. Ternate working for ShopCo, as long as he considers safety of the general public as paramount and he must be sure that his design includes all the required standards.

ASSOCIATION OF PROFESSIONAL ENGINEERS OF ONTARIO

PROFESSIONAL PRACTICE EXAMINARTION- December 11, 2004

PART "B" – Engineering Law and Professional Liability

This examination comes in two parts (**Part "A" and Part "B"**). <u>Both </u>parts must be completed in this sitting. You will be given a total of **180 minutes** to complete the examination.

Use the correct colour – coded Answer Book for each part, place in the correct envelope and **seal after completed.**

> *White Answer Book* for Part A white question paper.
> *Coloured Answer Book* for Part B coloured question paper.

This is a "**CLOSED BOOK**" examination. **No** aids are permitted other than the excerpts from the 1990 Ontario Regulation 941 covering sections 72 (*Professional Misconduct*) and 77 (*Code of Ethics*) supplied at the examination. Dictionaries are **not** permitted.

The marking of the questions will be based not only on academic content, but also on legibility and the ability to express yourself clearly and correctly in the English language. If you have any doubt about the meaning of a question, please state clearly how you have interpreted the question.

All **four** questions constitute a complete paper for Part "B". Each of the four questions is worth 25 marks.

Front Page

6.3

(MARKS)

(25)   1.   <u>Briefly</u> define and explain any <u>five</u> of the following:

   (i)   Gratuitous promise
   (ii)   Secret commission
   (iii)   Parol evidence rule
   (iv)   Statutory holdback
   (v)   Contract A
   (vi)   Vicarious liability
   (vii)   Common law
   (viii)   Undue influence

   2.   An owner / developer (the "owner") entered into a contract with an architectural firm (the "architect") for design and contract administration services in connection with the construction of a ten storey commercial office building.

   The building was designed to be entirely surrounded by a paved podium concrete deck used for parking and driving, and the design provided for a parking area below the deck. The podium deck was divided by construction joints and expansion joints placed to allow thermal expansion of the concrete as the temperature changed. The land on which the building was located sloped towards a river so the lower parking deck was designed to be partially open to the outside.

   The architect engaged a structural engineering firm (the "engineer"), as the architect's subconsultant on the project. The engineering firm, in its agreement with the architect, accepted responsibility for all structural aspects of construction, and also specifically acknowledged responsibility for the design of the paved podium concrete deck and the parking area below.

   Upon completion of the design and the tendering process, the owner entered into a contract for the construction of the project with an experienced contractor who had submitted the lowest bid.

   Unfortunately, within two years following construction, a significant number of leaks occurred in the podium deck which resulted in water leaks in the lower parking garage.

   The contract specifications had called for a specific rubberized membrane to be installed for the purpose of waterproofing the podium deck. However, during construction, at the suggestion of the roofing subcontractor and without the knowledge of the owner, another asphalt membrane product was substituted for the rubberized membrane product specified. Neither the engineer nor the architect objected to the substitution

when it was suggested. The roofing subcontractor had suggested the substitute membrane because it was more readily available and would speed completion of construction. The design engineer and the architect took the position that they would rely on the subcontractor's recommendation.

During the investigation into the cause of the leaks, another structural engineering firm provided its opinion that the rubberized membrane as specified in the contract was a superior product to the substituted membrane; that the substituted membrane was brittle and could fracture or crack under certain circumstances, particularly on podium decks with expansion joints; that the winter temperatures had contributed to the breakdown of the substitute membrane as it became more brittle at colder temperatures ; and that the substitute membrane should not have been used over expansion joints on a dynamic surface podium deck. The second engineering firm also expressed the opinion that the designers ought to have taken into account the non-static nature of the deck that featured these expansion joints and should not have accepted the substitute membrane.

Ultimately, to remedy the leaks, the substitute membrane had to be replaced by the rubberized membrane originally specified in the contract.

What potential liabilities in tort law arise in this case? In your answer, explain what principles of tort law are relevant and how each applies to the case.

(25)   **3.**   Clearwater Limited, a process-design and manufacturing company, entered into an equipment-supply contract with Pulverized Pulp Limited. Clearwater agreed to design, supply, and install a cleaning system at Pulverized Pulp's Ontario mill for a contract price of $800,000. The specifications for the cleaning system stated that he equipment was to remove ninety-eight percent of certain prescribed chemicals from the mill's liquid effluent in order to comply with the requirements of the environmental control authorities. However, the contract clearly provided that Clearwater accepted no responsibility whatsoever for any indirect of consequential damages, arising as a result of its performance of the contract.

The cleaning system installed by Clearwater did not meet the specifications, but this was not determined until after Clearwater had been paid $720,000 by Pulverizes Pulp. In fact, only seventy percent of the prescribed chemicals were removed from the effluent.

As a result, Pulverized Pulp Limited was fined $60,000 and was shut down by the environmental control authorities. Clearwater made several attempts to remedy the situation by altering the process and cleaning equipment, but without success.

2 of 3 pages   December 11, 2004. Part B - PPE

Pulverized Pulp eventually contracted with another equipment supplier. For an additional cost of $950,000, the second supplier successfully redesigned and installed remedial

process equipment that cleaned the effluent to the satisfaction of the environmental authorities, in accordance with the original contract specifications between Clearwater and Pulverized Pulp.

Explain and discuss what claim Pulverized Pulp Limited can make against Clearwater Limited in the circumstances. In answering, explain the approach taken by Canadian courts with respect to contracts that limit liability and include a brief summary of the development of relevant case precedents.

(25)   4.   A supplier of information technology hardware, ABC Hardware ("ABC"), submitted a fixed price bid on a computer installation project for a large accounting firm. ABC's bid price of six million dollars was very low in comparison to the other bidders. In fact, the three other bidders had each bid amounts in excess of nine million dollars.

The contract was awarded to the lowest bidder. The contract conditions expressly entitled the contractor to terminate the contract if the owner did not pay monthly invoices within thirty days following receipt of an invoice.

ABC commenced supplying computer hardware on the project and soon determined that it would likely suffer a major loss on the project, as it had made significant judgment errors in arriving at its bid price. ABC also learned that, in comparison with the other bidders, ABC had "left three million dollars on the table".

After the fifth invoice was delivered, ABC was approached by the accounting firm for additional information and explanation of bills from an equipment parts supplier, the cost which comprised a portion of the fifth invoice amount. The accounting firm requested that he additional information be provided prior to payment of the fifth invoice being due. Although the signed contract did not obligate ABC to obtain such additional information, a representative of ABC verbally informed the accounting firm that ABC would provide the additional information. However, ABC never did so.

Thirty-one days after the fifth invoice had been received, ABC notified the accounting firm that ABC was terminating the contract as the accounting firm had defaulted in its payment obligations under the specific wording of the contract. Was ABC entitled to terminate the contract? Explain the relevant legal principle and how it would be applied in this situation.

6.4 *Suggestive answers to Law, DECEMBER 2004*

Answer to question 1

(i)   It is a promise without consideration. A promise or a contract without consideration cannot be enforced. An exception exists in the case of equitable estoppel .

(ii)   It is money, or something of value that has been offered through a secret agreement to show a favour or disfavour to a third party. Secret commission is an offence under Criminal Code of Canada. If a Professional Engineer is involved in secret commission, then he/she will be charged under Regulation 941/90.

(iii)   "Parol" means verbal. It is a rule, which states that all the terms and conditions, which are mutually agreed upon must be mentioned in the contract, otherwise they cannot be enforced. This rule precludes evidence of omitted terms.

(iv)   This is a term, which is commonly used in construction contract. It is a percentage of money held back by the owner from the contractor. In Ontario it is 10% of the total price.
There are two types of holdback, 1-basic holdback or substantial performance, 2-finishing holdback. The owner is obligated to retain the holdback until he receives a notice that claims for lien related holdback is satisfied, expired or discharged. If the owner releases the holdback money too soon, then he may have to pay extra fund to satisfy the lien claimants.

(v)   There are two types of contracts in construction contracts. Contract A and contract B. When a bidder submits his bid in response to a call for bid, then contract A is formed. Contract A is an irrevocable contract and therefore must be submitted with a bond, a consideration.

(vi)   Vicarious means acting on behalf of somebody else. When an owner of a property sustains damages due to negligent conduct of an employee of a firm, then his/her employer would be vicariously liable for the damages.

(vii)   When a court decides about a case, it applies the legal principles made in the previous court cases. Such decision is known as "Common Law". It is a judge-made law. Common Law is used by all the provinces, but except Quebec.

(viii) It is similar to duress. It is when a person dominates the free will of another person to the extent to induce him/her to an unfair contract. Such as, relation between wife and husband, parent and a young child. Such a contract is not enforceable, and it is therefore void.

Answer to question 2

The principle of tort, which means injury, is to compensate a person who has sustained injury as a result of negligent conduct of the other person. In order to be successful in tort claim against the second person, there is no need for a contract to exist between them. The plaintiff must prove that:

i-   the defendant owed the plaintiff a duty of care
ii-   the defendant breached that duty by his conduct
iii-   the defendant negligent conduct caused injury to the plaintiff

In the given case the plaintiff is the "owner" and the defendants are "roofing subcontractor, engineer and the architect"

The contract had clearly specified that a specific rubberized membrane must be used.
Professional engineers are always duty bound to perform their profession with full competence and diligence. They should never approve anything without a thorough investigation. Both, the engineer and the architect accepted the suggested membrane without a thorough investigation. They had a duty of care for the owner. They knew or ought to have known that the owner was relying on them. They therefore breached the duty of care that they had for the owner. As a result of their negligence conduct a significant number of leaks occurred in the podium deck and ultimately in the lower deck. The liability of tort arises whenever the damage manifests itself. Since it is a tort case, there is no need that the default party must be in direct contract with the plaintiff. Consequently, the subcontractor, the engineer and the architect are liable to the owner. Such a case when more than one person is liable is known as concurrent tortfeasors.

Answer to question 3

This case is dealing with indirect or consequential damage. Indirect damage occurs when a contractor working on property of an owner accidentally damages something that constitutes damages, even economic loss, to the adjacent land or firm. For example, he may accidentally damage a power line of an adjacent firm resulting in shut down of the firm until the power supply is restored. In some occasion the owner may be fined due to contractor's non-compliance with the environmental protection.

At one time Canadian courts followed English courts precedent in applying what was called the "fundamental breach doctrine". According to this doctrine, an exemption clause cannot be enforced in case of a fundamental breach. Eventually, this doctrine was overruled by English courts. Even though, the concept of this doctrine is not completely overruled by Canadian courts, but strong preferences are given to the exemption clauses in commercial cases. Canadian courts following English courts do not disregard the exemption clauses. As a matter of fact parties to a contract are free to select and choose their own terms and conditions, provided that they follow the five essential terms making a contract binding. In the given case, Clearwater Limited is disclaiming any responsibility for any indirect damages as a result of its contractual performance. Since the court will honour the exemption clause disclaiming the responsibility, Clearwater would be protected against any liability.

Answer to question 4

This is the case, which involves "equitable estoppel", which states that when a party to a contract makes a gratuitous promise, a promise without consideration, and the second party to the contract reasonably relies on it, then it would be unfair and inequitable to allow the first party to revert to his contractual terms. Court always enforces only those terms and conditions that are clearly and unambiguously written in a contract and a contract without consideration cannot be enforced. If a term has verbally been agreed upon, but has not been entered into the contract, then this term is not enforceable. But in case of a gratuitous promise, since it would be unfair and inequitable to let the first party return to his strict contractual term, the concept of equitable estoppel is well applicable.

In the given case, a representative of ABC made a gratuitous promise, when the representative verbally informed the accounting firm that ABC would provide the additional information. In this case the concept of equitable estopple is well applicable and ABC must be estopped. It would be unfair and inequitable to allow ABC to return to his strict contractual term.

CHAPTER 7

PPE, Part "A", Part "B", 2005

Abstract. This chapter comprises only question papers of both parts, Part "A"- Professional Practice and Ethics, and Part "B" – Engineering Law and Professional Liability of August 2005 and Part "B" of December 2005.

ASSOCIATION OF PROFESSIONAL ENGINEERS OF ONTARIO

PROFESSIONAL PRACTICE EXAMINARTION- August 13, 2005

PART "A" – Professional Practice and Ethics

This examination comes in two parts **(Part "A" and Part "B")**. <u>Both </u>parts must be completed in this sitting. You will be given a total of **180 minutes** to complete the examination.

Use the correct colour – coded Answer Book for each part, place in the correct envelope and **seal after completed.**

> *White Answer Book for Part A white question paper:*
> *Coloured Answer Book for Part B coloured question paper:*

This is a **"CLOSED BOOK"** examination. **No** aids are permitted other than the excerpts from the 1990 Ontario Regulation 941 covering sections 72 (*Professional Misconduct*) and 77 (*Code of Ethics*) supplied at the examination. Dictionaries are **not** permitted.

The marking of the questions will be based not only on academic content, but also on legibility and the ability to express yourself clearly and correctly in the English language. If you have any doubt about the meaning of a question, please state clearly how you have interpreted the question.

All **four** questions constitute a complete paper for Part "A". Each of the four questions is worth 25 marks.

Where a question asks if a certain action by an engineer was ethical or not, a simple "yes" or "no" answer is not sufficient. You should identify the applicable clauses in Regulation 941 and comment on the action in each situation.

7.1

Question 1

(10)   (a)   Recently, the Professional Engineers Act and its Regulations were amended to add a new type of licence called the Provisional Licence. What is the purpose of the Provisional Licence and how is a Provincial Licence different from the "full" Licence given to a Member of the PEO?

(5)   (b)   Briefly define the term "profession".

(5)   ©   What is the "Fees Mediation Committee"? Describe its function.

(5)   (d)   What is the principal object of the Association of Professional Engineers of Ontario?

Question 2

Beacon Radio Manufacturing Inc. ("Beacon") is a well-known manufacturer of radio broadcast equipment. Beacon employs Hawker, a professional engineer, on a full- time basis as an engineering sales representative in Ontario.

In addition to working for Beacon, Hawker performs consulting professional engineering services for radio broadcasters. Hawker's services include technical analyses of broadcast systems. As part of these services, Hawker often makes recommendations for selection and purchase of broadcast equipment by broadcasters. Sometimes Hawker recommends Beacon's equipment.

Specify and explain:

(10)   (a)   the requirements, if any, stipulated in the Professional Engineers Act and its Regulations that Hawker must satisfy in order for Hawker to be allowed to provide such consulting professional engineering services; and

(15)   (b)   the requirements, if any stipulated in the Professional Engineers Act and its Regulations that Hawker must satisfy with respect to the manner in which such consulting professional engineering services are performed.

Question 3

AgriFab is a designer and manufacturer of farming equipment.

Recently, Farmer was seriously injured while operating a tractor designed and manufactured by AgriFab. In a letter to AgriFab, Farmer's lawyer claimed that the injury was due to a malfunction caused by a design error by AgriFab's engineering department. The letter threatened that AgriFab would be sued on account of Farmer's injuries.

AgriFab retains you (a Consulting Engineer) as an expert. Your services would be to investigate the failure and to give AgriFab your expert opinion on the cause of the failure. If the case goes to court, you could be called to testify as AgriFab expert witness. For your services AgriFab would pay you at an hourly rate. If you are called to testify in court and AgriFab wins the case, AgriFab would pay you a bonus in addition to your hourly rate.

Following your investigation, you conclude that the tractor was not designed properly and that Farmer was injured when certain safety feature of the tractor failed to function. You also conclude that it is likely that other farmers could be seriously injured while operating the particular tractor model You report your conclusions to AgriFab.

Based on your report, AgriFab promptly agrees to pay Framer $1 million. In exchange for the payment, Farmer agreed to give up the lawsuit and agreed to keep the payment a secret. The secrecy agreement was very important to AgriFab because AgriFab did not want future tractor sales to suffer from bad publicity.

AgriFab thanks you for your services and pays your fee.

(20)   (a)   Is there anything else you should do?

(5)   (b)   Please comment on the appropriateness of the fee structure according to which you would be paid.

Question 4

Elixir Engineering Ltd. ("Elixir") was engaged by Potion Pharmaceuticals Inc. ("Potion") to design a new production line at its pharmaceutical plant in Littletown, Ontario. In addition to preparing the design, Elixir would provide professional services during the installation of the project.

You are a professional engineer employed by Elixir The president of Elixir (who is also a professional engineer) has been very impressed with your work over the years and has personally requested that you be assigned to Potion's project to act as project manager. This is a very important job for Elixir and the assignment represents an important career opportunity for you (rumours have it that a successful project would result in your being promoted to vice president)

The completed design of the new production line includes a sophisticated new proprietary quality control system (designed and manufactured by Calibre Systems Corp. ("Calibre")). The system employs laser coding and machine vision to accurately identify, and label injection bottles containing various drugs that will be produced at the new facility. The technology uses a laser to code the cap of each injection bottle immediately after it is filled. After the bottles are sterilized, filled and transported through the system, a scanner reads the laser codes on the bottle caps and the appropriate labels are automatically printed and applied to the bottles. With the system, wrongly labeled bottles would be eliminated.

The project has progressed well(both on schedule and on budget) and is ready for start up, commissioning and testing. Unfortunately, during testing, the quality control system malfunctions and incorrectly labels 20% of the bottles. Calibre's technical personnel at the site are able to fix a number of problems contributing to the malfunction, and the error rate is reduced to 3 %. Calibre advises you that it will be unable to eliminate the remaining error without redesigning, manufacturing and testing a scanning component, a process which is expected to take at least two months.

You e-mail the bad news about the delay to Potion. The next day, Potion's project representative sends you an e-mail instruction that the production line be accepted with the 3% error rate. Potion's project representative explains that Potion is planning to use the new production line to manufacture a new patented drug, which will benefit thousands of people. Given the immediate demand for the drug, a two-month delay is not acceptable. The representative also indicates that Potion's and Elixir's presidents have already discussed the problem and have agreed that Potion would assume responsibility by legally indemnifying Elixir for any liability that could arise from the incorrect labeling.

(10) (a) Discuss your duties to the public. How do you evaluate the impact on the public interest? Does the drug's potential benefit to thousand of people affect your duties?

(5) (b) Do you have any duties to Potion and to your employer? What are they?

(5) ©   Are you able to fulfill all of your duties above? If not, what should you do?

(5) (d) What are the potential consequences if you do not fulfill your duties? Does Potion's promise to assume responsibility change things?

Any similarity in the question to actual person or circumstances is coincidental.

ASSOCIATION OF PROFESSIONAL ENGINEERS OF ONTARIO

PROFESSIONAL PRACTICE EXAMINARTION- August 13, 2005

PART "B" – Engineering Law and Professional Liability

This examination comes in two parts (**Part "A" and Part "B"**). Both parts must be completed in this sitting. You will be given **180 minutes** to complete the examination.

Use the correct colour – coded Answer Book for each part, place in the correct envelope and **seal after completed.**

> *White Answer Book for Part A white question paper.*
> *Coloured Answer Book for Part B coloured question paper.*

This is a "**CLOSED BOOK**" examination. **No** aids are permitted other than the excerpts from the 1990 Ontario Regulation 941 covering sections 72 (*Professional Misconduct*) and 77 (*Code of Ethics*) supplied at the examination. Dictionaries are **not** permitted.

The marking of the questions will be based not only on academic content, but also on legibility and the ability to express yourself clearly and correctly in the English language. If you have any doubt about the meaning of a question, please state clearly how you have interpreted the question.

All **four** questions constitute a complete paper for Part "B". Each of the four questions is worth 25 marks.

Front Page

7.2

(MARKS)

(25)   1. <u>Briefly</u> define and explain any <u>five</u> of the following:

    (i)   Discoverability concept.
    (ii)   Five examples of inappropriate conduct in the workplace prohibited by the
            Ontario Human Rights Code (it is not necessary to define these – just list 5
            examples)
    (iii)   Vicarious liability
    (iv)   Liquidated damages
    (v)   Contract A
    (vi)   Parol evidence rule
    (vii)   Secret commission

(25)   2.   National Stores Inc. ("NATIONAL"), the owner of a grocery store chain in
Ontario, contracted with an architect to design and prepare the construction
documentation for a new store in a town in northern Ontario.

        The architect produced some general construction specifications that included a
requirement that an automatic sprinkler system, conforming to the National Fire
Protection Association ("NFPA") standards, be installed.

        The architect retained an engineering firm pursuant to a separate agreement to
which NATIONAL was not a party. Under the contract the engineering firm was to
prepare the detailed engineering design for the project, including the sprinkler system.
The engineering design was to conform to the architect's general specifications.

        A recent engineering graduate employed by the engineering firm prepared the
design of the sprinkler system. Not being familiar with the NFPA requirements, the
employee read certain sections of the standards but did not have enough time, given
other project responsibilities, to pay close attention to all the details. A professional
engineer reviewed the employee's completed sprinkler system design. Although the
professional engineer did not perform a detailed check, the professional engineer
considered the design satisfactory.

        Six months after the store opened for business, a fire occurred early one
morning. The fire caused substantial damage to the store and to its inventory and
NATIONAL had to close the store for repair.

NATIONAL retained a consulting engineer to conduct an independent investigation. The consulting engineer determined that the sprinkler system was inadequately designed. Specifically, the design did not conform to the NFPA standards, which required, among other things that the coverage per sprinkler head was not to exceed 10 square meters. The engineer determined that 10 percent of the sprinkler heads were designed to cover an area as high as 25 square meters. The report indicated that, in the engineer's expert opinion, had the sprinkler head spacing conformed to the NFPA standards, the fire should have been quickly extinguished and would not have spread to any great extent.

What liabilities in *tort law* may arise in this case? In your answer, explain the purpose of tort law and identify what essential principles of tort law are relevant. Apply each principle to the facts. Indicate a likely outcome of the matter.

(25)   3.   A manufacturing company (the "Environmental Supplier"), supplying environmental control equipment to industry, entered into an equipment supply contract with a mining company in northern Ontario. The Environmental Supplier agreed to design, supply, and install a dust collection system at the mining company's northern Ontario smelter for a contract price of $900,000. The specifications for the dust collection system stated that the dust collection equipment was to remove 95 percent of prescribed exhaust particles from the exhaust gases in order to comply with applicable statutory environmental control requirements.

In addition, the signed contract between the Environmental Supplier and the mining company also contained provision limiting to $900,000, the Environmental Supplier's total liability for any loss, damage, or injury resulting from its performance of design, supply, and installation services to the mining company pursuant tot the contract.

The dust collection system as installed by the Environmental Supplier did not meet the specifications. In fact, only 60 percent of the prescribed exhaust particles were removed from the exhaust gases. As a result, the mining company was faced with the threat of substantial fines and possible shutdown by environmental control authorities. The Environmental Supplier refused to remedy the defective equipment without being assured of compensation from the mining company of any costs in excess of $900,000 incurred in connection with such remedial work.

At the time of discovering that the system failed to meet the specifications, the Environmental Supplier had already received $600,000 from the mining company, and the mining company refused to pay anything further to the Environmental Supplier

The mining company contacted another equipment supplier, who, for an additional cost of $1,400,000 successfully designed and installed remedial equipment sufficient to clean the exhaust gases to the satisfaction of the environmental authorities and in accordance with the original contract specifications between the Environmental Supplier and the mining company.

Explain and discuss what claim the mining company can make against the Environmental Supplier in the circumstances. In answering, please include a summary of the development of relevant case precedents.

(25) **4.**   An information technology firm submitted a bid to design and install software and hardware for an electronic technology process to control the operation of a large scale sorting equipment for a major international courier company.

The firm's fixed guaranteed maximum price was the lowest bid and the contract was awarded to it. The contract condition entitled the information technology firm to terminate the contract if the courier company did not pay monthly progress payments within 15 days following certification that a progress payment was due. Pursuant to the contract, the certification was carried out by an independent engineering firm engaged as contract administrator.

The work under the contract was to be performed over a 5 month period. After commencing work on the project the information technology firm determined that it had made significant judgment errors in arriving at its bid price and that it would face a major loss on the project. Its concern about the anticipated loss was increased further when it also leaned that, in comparison with the other bidders, its bid price was extremely low and that, in winning the bid, it had left more than one million dollars "on the table".

Two monthly progress payments were certified as due by the independent engineering firm and paid by the courier company in accordance with the terms of the contract. However, after the third monthly progress payment was certified as due by the independent engineering firm, the courier company's finance department asked the information technology firm's representative on the project for additional information relating to an invoice from a subcontractor to the information technology firm. The subcontractor's invoice comprised a portion of the third progress payment amount. The courier company's finance department requested that the additional information be provided prior to payment of the third progress payment.

There was nothing in the signed contract between the information technology firm and the courier company that obligated the information technology firm to provide the additional information on the invoice from its subcontractor. However, the information technology firm's representative did verbally indicate to the courier company's finance department that the additional information would be provided.

The additional information relating to the subcontractor's invoice was never provided by the information technology firm

Sixteen days after the third progress payment had been certified for payment, the information technology firm notified the courier company in writing that it was terminating the contract because the courier company was in default of its obligations to make payments within fifteen days pursuant to the express wording of the contract.

Was information technology firm entitled to terminate the contract in these circumstances? In giving reasons for your answer, identify and explain the relevant legal principle and how it would apply.

ASSOCIATION OF PROFESSIONAL ENGINEERS OF ONTARIO

PROFESSIONAL PRACTICE EXAMINARTION- December 17, 2005

PART "B" – Engineering Law and Professional Liability

You will be given **90 minutes** to complete this examination.

Use the correct colour – coded Answer Book for each part, place in the correct envelope and seal after completed.

> *White Answer Book* for Part A white question paper.

> *Coloured Answer Book* for Part B coloured question paper.

This is a "**CLOSED BOOK**" examination. **No** aids are permitted other than the excerpts from the 1990 Ontario Regulation 941 covering sections 72 (*Professional Misconduct*) and 77 (*Code of Ethics*) supplied at the examination. Dictionaries are not permitted.

The marking of the questions will be based not only on academic content, but also on legibility and the ability to express yourself clearly and correctly in the English language. If you have any doubt about the meaning of a question, please state clearly how you have interpreted the question.

All **four** questions constitute a complete paper for Part "B". Each of the four questions is worth 25 marks.

Front Page

7.3
(MARKS)

(25)   1. <u>Briefly</u> define and explain any <u>five</u> of the following:

      (i)   Rule of contra proferetem
      (ii) The limitation period applicable to contracts in Ontario
      (iii) Five examples of inappropriate conduct in the workplace (list only)
      (iv) Contract A (in tendering)
      (v)   Statutory holdback under the *Construction Lien Act*
      (vi) Discharge by frustration
      (vii) The corporate director's standard of care
      (viii) Quantum meruit

(25)   **2.**   A manufacturing company retained an architect to design a new plant. The manufacturer, as client, and the architect entered into a written client/architect agreement in connection with the project. The purpose of the plant construction was to enable the client to expand its manufacturing and warehousing facilities.

    The structural design of the plant was prepared by an engineering firm which was retained by the architect. A separate agreement was entered into between the architect and the engineering firm to which the client was not a party.

    The engineering firm turned the matter over to one of its employees, a professional engineering with experience in structural steel design who proceeded to complete the structural design of the plant. The client had informed the architect that the second floor of the plant was to be used for manufacturing and warehousing purposes and that forklift trucks would be extensively used in both the manufacturing and warehousing sections on the second floor. The architect passed this information on to the engineering firm. The employee engineer designed a steel frame and specified that the second floor was to be a concrete –steel composite, consisting essentially of concrete poured onto a steel deck, and containing a light steel mesh. The steel deck, concrete thickness and steel mesh specifications were specified in the Engineer's design and were taken from design tables which the engineer located in his firm's library and which had been published by a company which manufactured and supplied the steel deck.

    The construction of the plant was completed and shortly after manufacturing commenced at the plant, severe cracks appeared in the concrete on the second floor. After two months of operation the floor cracked and broke up so badly that the plant had to be shut down and a remedial floor slab, heavily reinforced with reinforcing bar, was poured on top of the damaged second floor.

The design of the remedial floor slab was carried out by another consulting engineering firm. After completing its investigation of the cause of the failure of the second floor, the second engineering firm stated that, in its opinion the engineer who had designed the second floor had used design tables from steel deck manufacturer which was 12 years out of date and had also failed to use the tables that the engineer obviously ought to have used knowing that the floor was intended for manufacturing and forklift truck loading. The second consulting engineering firm concluded that the depth of concrete and size of steel mash on the floor as initially designed resulted in a floor that might have been appropriate for the design of an office or apartment building but not for manufacturing and warehousing purposes.

What potential liabilities in tort law arise from the preceding set of facts? In your answer, state the essential principles applicable in a tort action and apply these principles to the facts. Indicate a likely outcome of the matter.

(25)   **3.**   A $30,000,000 contract for the design, supply and installation of a cogeneration facility was entered into between a pulp and paper company("Pulpco") and an industrial contractor. The cogeneration facility, the major components of which included a gas turbine, a heat recovery steam generator and a steam turbine, was to be designed and constructed to simultaneously generate both electricity and steam for use by Pulpco in its operations.

The contract provided that the electrical o\power generated by the cogeneration facility was not to be less than 25 megawatts. A liquidated damages provision was included in the contract specifying a pre-estimated amount payable by the contractor to Pulpco for ach megawatt of electrical power generated less than the minimum 25 megawatts specified. Other provisions specified additional liquidated damages at prescribed rates relating to other matters under the contract, including any failure by the contractor to meet the required heat rates or to achieve completion of the facility for commercial use by a stipulated date. However, the contract also included a maximum liability" provision that limited to $5,000,000 the contractor's liability for all liquidated damages due to failure to achieve (i) the specified electrical power output, (ii) the guaranteed heat rate and (iii) the specified completion date. The contract clearly provided that under no circumstances was the contractor to be liable for any damages beyond the overall total of $5,000,000 for liquidated damages. Pulpco's sole and exclusive remedy for damages under the contract was strictly limited to the total liquidated damages, up to the maximum of $5,000,000. The contract specified that Pulpco was not entitled to make any other claim for damages, whether on account of any direct, indirect, special or consequential damages, howsoever caused.

Unfortunately the contractor's installation fell far short of the electrical power generation specifications (achieving less than 25% of the specified megawatts) and

2 of 4 pages   December 17, 2005, Part B – PPE

the heat rate specifications provided in the contract. The contractor was paid $27,000,000 before the problems were identified on startup and testing. Because of its very poor performance, the contractor also failed to meet the completion date by a very substantial margin. Applying the liquidated damages provisions, the contractor's overall liability for all liquidated damages under the contract totaled $4,000,000. Ultimately, Pulpco had to make arrangements through another contractor for new equipment items and parts to be ordered and installed in order to enable the cogeneration facility to meet the technical specifications, with the result that the total cost of the replacement equipment and parts reached an additional $15,000,000 beyond the original contract price of $30,000,000.

Explain and discuss what claim Pulpco could make against the contractor in the circumstances. In answering, explain the approach taken by Canadian courts with respect to contracts that limit liability and include a brief summary of the development of relevant case precedents.

(25)   4.   A mining contractor signed an option contract with a land owner which provided that if the mining contractor (the "optionee") performed a specified minimum amount of exploration services on the property of the owner (the "optionor") within a nine month period, then the optionee would be entitled to exercise its option to acquire certain mining claim from the optionor.

Before the expiry of this nine-month "option period", the optionee realized that it couldn't fulfill its obligation to expend the required minimum amount by the expiry date. The optionee notified the optionor of its problem, prior to expiry of the option period and the optionor indicated that the option period would be extended. However, no written record of this extension was made, nor did the optionor receive anything from the optionee in for the extension.

The optionee then proceeded to perform the services and to finally expend the specified minimum amount during the extension period. However, when the optionee attempted to exercise its option to acquire the mining claims the optionor took the position that, on the basis of the strict wording of the signed contract, the optionee had not met its contractual obligations. The optionor refused to grant the mining claims to the optionee.

Was the optionor entitled to deny the optionee's exercise of the option? Identify the contract law principles that apply, and explain the basis of such principles and how they apply, to the positions taken by the optionor and by the optionee.

CHAPTER 8

Professional Engineers Act

Abstract. In this chapter the candidates are provided with definitions of by-laws, Power of Minister, regulations, executive committee, issuance of licence, issuance of certificate of authorization, supervision by professional engineer, duties of registrar, duties of complaints committee, registrar investigation, confidentiality, penalties and many more.

Definitions

1. In this Act,

"Academic Requirements Committee" means the academic requirements committee appointed pursuant to the regulations; ("comite de verification des dipl6mes")

"architect" means a person who is licensed or who holds a certificate of practice or a temporary licence under the Architects Act; ("architecte")

"Association" means the Association of Professional Engineers of Ontario; ("Ordre")

"by-laws" means by-laws made under this Act; ("reglements administratifs")

"certificate of authorization" means a certificate of authorization issued under this Act to engage in the business of providing services that are within the practice of professional engineering; ("certificat d'autorisation")

"Complaints Review Councillor" means the complaints review councillor appointed under this Act; ("conseiller mediateur")

"Council" means the Council of the Association; ("Conseil")

"Experience Requirements Committee" means the experience requirements committee appointed pursuant to the regulations; ("comite de verification de l'experience")

"Joint Practice Board" means the Joint Practice Board established under the Architects Act; ("Conseil professionnel mixte")

"licence" means a licence to engage in the practice of professional engineering issued under this Act; ("permis")

"limited licence" means a limited licence to engage in the practice of professional engineering issued under this Act; ("permis restreint")

"Minister" means the Attorney General or such other member of the Executive Council as is designated by the Lieutenant Governor in Council; ("ministre")

"Practice of professional engineering" means any act of designing, composing, evaluating, advising, reporting, directing or supervising wherein the safeguarding of life, health, property or the public welfare is concerned and that requires the application of engineering principles, but does not include practising as a natural scientist; ("exercice de la profession d'ingenieur") "professional engineer" means a person who holds a licence or a temporary licence; ("ingenieur")

"provisional licence" means a provisional licence to engage in the practice of professional engineering issued under subsection 14(7); ("permis provisoire")

"Registrar" means the Registrar of the Association; ("registra-teur")

"regulations" means the regulations made under this Act; ("reglements")

"temporary licence" means a temporary licence to engage in the practice of professional engineering issued under this Act. ("permis temporaire") RS.O. 1990, c. P.28, s. 1; 1999, c. 12, Sched. B, s. 13; 2001, c. 9, Schcd. B, s. 11(1).

Association

2.(1) The Association of Professional Engineers of the Province of Ontario, a body corporate, is continued as a corporation without share capital under the name of Association of Professional Engineers of Ontario in English and Ordre des ingenieurs de l'Ontario in French. R.S.O. 1990, c. P.28, s. 2(1).

Head office

(2) The head office of the Association shall be in the City of Toronto. RS.O. 1990, c. P.28, s. 2(2); 1997, c Sched. 26,

Principal object

(3) The principal object of the Association is to regulate the practice of professional engineering and to govern its members, holders of certificates of authorizations, holders of temporary licences, holders of provisional licences and holders of limited licences in accordance with this Act, the regulations and the by-laws in order that the public interest may be served and protected. RS.O. IS 00, c. P.28, s. 2(3); 2001, c. 9, Sched. B, s. 11(2).

Additional objects

(4) For the purpose of carrying out its principal object, the Association has the following additional objects

1. To establish, maintain and develop standards of knowledge and skill among its members.

2. To establish, maintain and develop standards of qualification and standards of practice for the practice of professional engineering.

3. To establish, maintain and develop standards of professional ethics among its members.

4. To promote public awareness of the role of the Association.

5. To perform such other duties and exercise such other powers as are imposed or conferred on the Association by or under any Act. R.S.O. 1990, c., P.28, s. 2(4).

Capacity and powers of Association

(5) For the purpose of carrying out its objects, the Association has the capacity and the powers of a natural person. RS.O. 1990, c. P 28,s. 2(5).

Council of Association

3.(1) The Council of the Association is continued and shall be the governing body and board of directors of the Association and shall manage and administer its affair R.S.O.1990, c. P. 28,s. 3(1).

Composition of Council

(2) The Council shall be composed of,

(a) not fewer than fifteen and not more than twenty persons who are members of the Association and who are elected by the members of the Association as provided by the regulations;

(b) not fewer than five and not more than seven persons who are members of the Association and who are appointed by the Lieutenant Governor in Council;

(c) not fewer than three and not more than five persons who are not members of the governing body of a self-regulating licensing body under any other Act or licensed under this Act and who are appointed by the Lieutenant Governor in Council; and

(d) the holders of offices prescribed by the regulations who are not members of the Council under clause (a), (b) or (c). R.S.O. 1990, c. P.28, s. 3(2).

Idem

(3) No person shall be elected or appointed to the Council unless he or she is a Canadian citizen resident in Ontario. R.S.O. 1990, c. P.28, s. 3(3)

Remuneration of lay members

(4) The persons appointed under clause (2)(c) shall be paid, out of the money appropriated therefore by the Legisl- ature, such expenses and remuneration as is determined by the Lieutenant Governor in Council. R.S.O. 1990, c. P.28, s. 3(4).

Term of office of appointed members

(5) In each year, the persons to be appointed by the Lieutenant Governor in Council shall be appointed for one year, two year or three year terms in order that one-third, or as near thereto as possible, shall be appointed in each year. R.S.O. 1990, c. P.28, s. 3(5).

Qualifications to vote

(6) Every member of the Association who is not in default of payment of an annual fee prescribed by the by-laws is qualified to vote at an election of members of the Council. R.S.O. 1990, c. P.28, s. 3(6).

Officers

(7) The Association shall have the officers provided for by the regulations. R.S.O. 1990, c. P. 28, s. 3(7).

Registrar and staff

(8) The Council shall appoint during pleasure a Registrar, who shall be a member of the Association, and may appoint one or more deputy registrars who shall have the powers of the Registrar for the purposes of this Act, and may appoint such other persons as are from time to time necessary or desirable in the opinion of the Council to perform the work of the Association. R.S.O. 1990, c. P.28, s. 3(8); 2001, c. 9, Sched. B, s. 11(3).

Role of Registrar

(8.1) The Registrar is responsible for the administration of the Association and reports to the Council. 2001, c. 9, Sched. B,s. 11(4).

Quorum

(9) A majority of the members of the Council continues a quorum. R.S.O. 1990, c. P.28, s. 3(9).

Vacancies

(10) Where one or more vacancies occur in the membership of the Council, the

members remaining in office constitute the Council so long as their number is not fewer than a quorum. R.S.O. 1990, c. P.28, s. 3(10).

Filling of vacancy

(11) A vacancy on the Council caused by the death, resignation, removal or incapacity to act of an elected member of the Council shall be filled as soon as practicable by a member of the Association,

(a) where a quorum of the Council remains in office, appointed by the majority of the Council, and the member so appointed shall be deemed to be an elected member of the Council; or

(b) where no quorum of the Council remains in office, elected in accordance with the regulations; and the member so appointed or elected shall hold office for the unexpired portion of the term of office of the member whose office he or she is elected or appointed to fill. R.S.O. 1990, c. P.28, s. 3(11).

Meetings of Council

(12) The Council shall meet at least four times a yea. R. S. O.1990, c. P.28, s. 3(12).

Annual meetings

4. The Association shall hold an annual meeting of the members of the Association not more than fifteen months after the holding of the last preceding annual meeting. R.S.O. 1990, c. P.28, s. 4.

Membership

5.(1) Every person who holds a licence is a member of the Association subject to any term, condition or limitation to which the licence is subject.

Resignation of membership

(2) A member may resign his or her membership by filing with the Registrar a resignation in writing and his or her licence is thereupon cancelled, subject to the continuing jurisdiction of the Association in respect of any disciplinary action arising out of the person's professional conduct while a member. R.S.O. 1990, c. P. 28, s. 5

Powers of Minister

6. In addition to his or her other powers and duties under this Act, the Minister may,
(a) review the activities of the Council;

(b) request the Council to undertake activities that, in the opinion of the Minister, are necessary and advisable to carry out the intent of this Act;

(c) advise the Council with respect to the implementation of this Act and the regulations and with respect to the methods used or proposed to be used by the Council to implement policies and to enforce its regulations and procedures. R.S.O. 1990, c. P.28,s. 6.

Regulations

7.(1) Subject to the approval of the Lieutenant Governor in Council and with prior review by the Minister, the Council may make regulations,

1. fixing the number of members to be elected to the Council under clause 3(2)(a) and defining constituencies, and prescribing the number of representatives;

2. respecting and governing the qualifications, nomination, election and term or terms of office of the members to be elected to the Council, and controverted elections;

3. prescribing the conditions disqualifying elected members from sitting on the Council and governing the filling of vacancies on the Council;

4. prescribing positions of officers of the Association and providing for their election or appointment;

5. respecting the composition of the committees required by this Act, other than the Complaints Committee and the Discipline Committee, the mechanism of the appointment of members of the committees and procedures ancillary to those specified in this Act in respect of any committee;

6. respecting matters of practice and procedure before committees required under this Act that do not conflict with the Statutory Powers Procedure Act;

7. prescribing the quorums of the committees required by this Act other than the Complaints Committee and the Discipline Committee;

8. prescribing classes of persons whose interests are related to those of the Association and the privileges of members of the classes in relation to the Association;

9. respecting any matter ancillary to the provisions of this Act with regard to the issuing, suspension and revocation of licences, certificates of authorization, temporary licences, provisional licences and limited licences, including but not limited to regulations respecting,

i. the scope, standards and conduct of any examination set or approved by the Council as a licensing requirement,

ii. the curricula and standards of professional training programs offered by the Council,

iii. the academic, experience and other requirements for admission into professional training programs,

iv. classes of licences,

v. the academic and experience requirements for the issuance of a licence or any class of licence, and

vi. classes of certificates of authorization temporary licences, provisional licences and limited licences, including prescribing requirements and qualifications for the issuance of specified classes of certificates of authorization temporary licences, provisional licences and limited licences, and terms and conditions that shall apply to specified classes of certificates of authorization, temporary licences, provisional licences and limited licences;

10. prescribing forms of applications for licences, certificates of authorization, temporary licences provisional licences and limited licences and requiring their use;

11. requiring the making of returns of information in respect of the holdings of shares and the officers and directors of corporations that apply for or hold certificates of authorization and in respect of the interests of partners that apply for or hold certificate of authorization and prescribing and requiring the use of forms of such returns;

12. requiring and governing the signing and sealing of documents and designs by members of the Association, holders of temporary licences and holders of limited licences, specifying the forms of seals and

respecting the issuance and ownership of seals;

13. requiring the making of returns of information by members of the Association and holders of certificates of authorization, temporary licences provisional licences and limited licences in respect of names, addresses, telephone numbers, professional associates, partners, employees and professional liability insurance, and prescribing and requiring the use of forms of such returns;

14. requiring and governing the disclosure of the identity of holders of certificates of authorization on documents and designs involving the practice of professional engineering issued by such holders and specifying the form and manner of such disclosure;

15. governing the use of names and designations in the practice of professional engineering by members of the Association and holders of certificate of authorization, temporary licences, provisional Iicences and limited licences

16. providing for the maintenance and inspection of registers of members of the Association, holders of temporary licences, holders of limited licences, holders of provisional licences and holders of certificates of authorization;

17. prescribing and governing standards of practice and performance standards for the profession

18. providing for the setting of schedules of suggested fees for professional engineering services and for the publication of the schedules;

19. respecting the advertising of the practice of professional engineering;

20. prescribing a code of ethics;

21. defining professional misconduct for the purposes of this Act

22. providing for the designation of members of the Association and holders of temporary licences as specialists, prescribing the qualifications and requirements for designation as a specialist, providing for the suspension or revocation of such a designation and for the regulation and prohibition of the use of the designation by members of the Association a holder of a temporary licence or a certificate of authorization;

23. providing for the designation of members of the Association as consulting engineers, prescribing the qualifications and requirements for designation as a consulting engineer, providing for the suspension or revocation of such a designation and for the regulation and prohibition of the use of the designation by members of the Association, a holder of a temporary licence or a certificate of authorization;

24. prescribing the minimum requirements for professional liability insurance, requiring the delivery to the Registrar of proof of such insurance and prescribing the form of such proof and the manner and time of the delivery;

25. prescribing the amount of and requiring the payment of annual fees by holders of certificates of authorization, temporary, provisional and limited licences and by students and members of related classes recognized by the Association, and fees for temporary licences, provisional licences, limited licences, certification, registration, designations, examinations and continuing education, including penalties for late payment, and fees for anything the Registrar

is required or authorized to do, and prescribing the amounts thereof;

26. providing for the entering into of arrangements by the Association for its members and holders of certificates ofauthorization, temporary licences, provisional and limited licences respecting indemnity for professional liability and requiring the payment and remittance of premiums in connection therewith and prescribing levies to be paid by members and holders of certificates of authorization, temporary licences, provisional and limited licences in respect of such indemnity for professional liability;

27. providing for continuing education of members;

28. respecting the duties and authority of the Registrar;

29. prescribing qualifications and requirements that shall be complied with to obtain the reinstatement of a licence, certificate of authorization, temporary licence or limited licence that was cancelled by the Registrar;

30. classifying and exempting any class of holders of licences, certificates of authorization, temporary licences or limited licences from any provision of the regulations under such special circumstances in the public interest as the Council considers advisable;

31. exempting any act within the practice of professional engineering from the application of this Act

32. specifying acts within the practice of professional engineering that are exempt from the application of this Act when performed or provided by a member

of a prescribed class of persons, and prescribing classes of persons for the purpose of the exemption

33. despite anything else in this Act, providing for the payment of start-up funding to the Ontario Society of Professional Engineers during the three-year period that begins on the day the Red Tape Reduction Act, 2000 receives Royal Assent, and specifying the amounts to be paid, the time and manner of payment and the conditions to be met before each payment is made. R.S.O. 1990, c. P.28, s 7(1);2000.c. 26, Sched. A, s.12, 2001, c. 9,Sched.B, s. 11(5-12).

Distribution of regulations

(2) A copy of each regulation made under subsection (1)

(a) shall be forwarded to each member of the Association and to each holder of a certificate of authorization, temporary licence, provisional licence or limited licence; and

(b) shall be available for public inspection in the office of the Association. R.S.O. 1990, c. P.28, 7(2) 2001.c .9, Sched. B, s. 11(13).

By-laws

8.(1) The Council may pass by-laws relating to the administrative and domestic affairs of the Association not inconsistent with this Act and the regulations and, without limiting the generality of the foregoing,

1. prescribing the seal and other insignia of the Association and providing for their use;

2. providing for the execution of documents by the Association:

3. respecting banking and finance;

4. fixing the financial year of the Association and providing for the audit of the accounts and transactions of the Association;

5. respecting the calling, holding and conducting of meetings of the Council and the duties of members of Council;

6. providing for meetings of the Council and committees, except in a proceeding in respect of a membership, certificate of authorization, temporary licence, provisional licence or limited licence, by means of conference telephone or other communications equipment by means of which all persons participating in the meeting can hear each other and a member of the Council or committee participating in a meeting in accordance with such by- law shall be deemed to be present in person at the meeting

7. providing that the Council or a committee may act upon a resolution consented to by the signatures of all members of the Council or the committee except in a proceeding in respect of a licence, certificate of authorization, temporary licence, provi-sional licence or limited licence, and a resolution so consented to in accordance with such a by-law is as valid and effective as if passed at a meeting of the Council or the committee duly called, constituted and held for that purpose;

8. respecting the calling, holding and conducting of meetings of the membership of the Association;

9. authorizing voting by mail by the general membership of the Association on any of the business of the Association and prescribing procedures for such voting;

10. prescribing the duties of officers of the Association;

11. prescribing forms and providing for their use;

12. providing procedures for the making, amending and revoking of the by-laws;

13- respecting management of the property of the Association;

14. providing for the appointment, composition, powers, duties and quorums of additional or special committees;

15. respecting the application of the funds of the Association and the investment and reinvestment of any of its funds not immediately required, and for the safekeeping of its securities;

16. prescribing the amount and requiring the payment of annual fees by members of the Association;

17. respecting the borrowing of money by the Association and the giving of security therefore ;

18. respecting membership of the Association in other organizations the objects of which are not inconsistent with and are complementary to those of the Association, the payment of annual assessments and provision for representatives at meetings;

19. providing for the establishment and dissolution and governing the operation of groups of members of the Association and respecting grants by the Association to any such groups;

20. authorizing the making of grants for any purpose that may tend to advance

knowledge of professional engineering education, or maintain or improve the standards of practice in professional engineering or support and encourage public information and interest in the past and present role of professional engineering in society;

21. respecting scholarships, bursaries and prizes related to the study of professional engineering;

22. respecting the establishment and operation and use of publications of the Association;

23. providing for an employment advisory service and for the continuance of the retirement savings plans in which members of the Association may partici- pate on a voluntary basis;

24. regarding such other matters as are entail in carrying on the business of the Association and are not included in section 7. R.S.O. 1990, c. P.28, s. 8(1); 2001, c. 9, Sched. B. s. 11(14, 15).

When by-laws come into force

(2) A by-law passed by the Council is not effective until confirmed by the members of the Association. R.S.O. 1990,c. P.28, s. 8(2).

Confirmation of by-laws

(3) A by-law passed by the Council may be confirmed by the members of the Association only by means of a vote conducted by mail. R.S.O. 1990, c. P.28. s. 8(3).

Distribution of by-laws

(4) A copy of the by-laws made under subsection (1) and amendments thereto,

(a) shall be forwarded to the Minister;

(b) shall be forwarded to each member of the Association; and

(c) shall be available for public inspection in the office of the Association. R.S.O. 1990, c. P.28, s 8(4).

Official publication

9. The Council shall establish and designate an official publication of the Association. R.S.O. 1990, c. P. 28, s .9.

Establishment of committees

10.(1) The Council shall establish and appoint the following committees:

(a) Executive Committee;

(b) Academic Requirements Committee;

(c) Experience Requirements Committee;

(d) Registration Committee;

(e) Complaints Committee;

(f) Discipline Committee;

(g) Fees Mediation Committee,

and may establish such other committees as the Council from time to time considers necessary.

Vacancies

(2) Where one or more vacancies occur in the membership of a committee, the members remaining in office constitute the committee so long as their number is not fewer than the prescribed quorum. R.S.O. 1990, c. P.28, s. 10.

Executive Committee

11. The Council may delegate to the Executive Committee the authority to exercise any power or perform any duty of the Council other than to make, amend or revoke a regulation or a by-law. R.S.O. 1990, c. P.28, s 11.

When licences or certificates required
Licensing requirement

12.(1) No person shall engage in the practice of professional engineering or hold himself, herself or itself out as engaging in the practice of professional engineering unless the person is the holder of a licence, a temporary licence, a provisional licence or a limited licence. R.S.O.1990, c. P.28, s. 12(1); 2001, c. 9, Sched. B, s. 11(16).

Certificate of authorization

(2) No person shall offer to the public or engage in the business of providing to the public services that are within the practice of professional engineering except under and in accordance with a certificate of authorization. R.S.O.1990, c. P.28, s. 12(2).

Exceptions

(3) Subsections (1) and (2)-do not apply to prevent a person,

(a) from doing an act that is within the practice of professional engineering in relation to machinery or equipment, other than equipment of a structural nature, for use in the facilities of the person's employer in the production of products by the person's employer;

(b) from doing an act that is within the practice of professional engineering where a professional engineer assumes responsibility for the services within the

practice of professional engineering to which the act is related;

(c) from designing or providing tools and dies;

(d) from doing an act that is within the practice of professional engineering but that is exempt from the application of this Act when performed or provided by a member of a class of persons prescribed by the regulations for the purpose of the exemption, if the person is a member of the class;

(e) from doing an act that is exempt by the regulations from the application of this Act;

(f) from using the title "engineer" or an abbreviation of that title in a manner that is authorized or required by an Act or regulation. R.S.O. 1990, c. P.28, s. 12 (3); 2001, c. 9, Sched. B, s. 11(17).

Idem

(4) Subsections (1) and (2) do not apply to the preparation or provision of a design for the construction, enlargement or alteration of a building,

(a) that is not more than three storeys and not more than 600 square metres in gross area as constructed, enlarged or altered;

(b) that is used or intended for one or more of residential occupancy, business occupancy, personal services occupancy, mercantile occupancy or industrial occupancy; and

(c) is not designed to house and is not part of an apparatus, process, facility or works the design of which is within the practice of professional engineering. R.S.O. 1990, c. P.28, s. 12(4).

Idem

(5) Subsections (1) and (2) do not apply to,

(a) the preparation or provision of a design for the construction, enlargement or alteration of a building that is not more than three storeys and that is used or intended for residential occupancy and

(i) that contains one dwelling unit or two attached dwelling units each of which is constructed directly on grade, or

(ii) that is not more than 600 square metres in building area as constructed, enlarged or altered and contains three or more attached dwelling units, each of which is constructed directly on grade, with no dwelling unit constructed above another dwelling unit; or

(b) the preparation or provision of a design for alterations within a dwelling unit that will not affect or are not likely to affect fire separations, firewalls, the strength or safety of the building or the safety of persons in the building. R.S.O. 1990, c. R28, 12(5).

Idem

(6) The following rules govern the relationship between professional engineers and architects and subsection (1) and (2) do not apply to prevent an architect from preparing or providing a design for and carrying out the general review of the construction, enlargement or alteration of a building in accordance with these rules:

1. Only an architect may prepare or provide design for the construction, enlargement or alteration of a building,

i. used or intended for residential occupancy,

ii. that exceeds 600 square metres in gross area, and

iii. that does not exceed three storeys, and carry out the general review of the construction, enlargement or alteration of the building but an architect who prepares or provides such a design may engage a professional engineer to provide services within the practice of professional engineering in connection with the design and the professional engineer may provide the services.

2. A professional engineer or an architect may prepare or provide a design for the construction, enlargement or alteration of a building,

i. that exceeds 600 square metres in gross area or three storeys, and

ii. that is used or intended for,

   A. industrial occupancy, or

   B. mixed occupancy consisting of industrial occupancy and one or more other occupancies, where none of the other occupancies exceeds 600 square metres of the gross area, but only a professional engineer may provide services within the practice of professional engineering in connection with the design.

3. Subject to rules 4 and 5, a professional engineer shall provide services that are within the practice of professional engineering and an architect shall provide services that are within the practice of architecture related to the construction, enlargement or alteration of a building used or intended for,

i. assembly occupancy,

ii. institutional occupancy,

iii. business occupancy or personal services occupancy that exceeds 600 square metres in gross area or three storeys,

iv. mercantile occupancy that exceeds 600 square metres in gross area or three storeys,

v. residential occupancy that exceeds three storeys,

vi. mixed occupancy consisting of industrial occupancy and one or more other occupancies, where one of the other occupancies exceeds 600 square metres in gross area,

vii. mixed occupancy consisting of a combination of,

A. assembly occupancy and any other occupancy, except industrial occupancy,

B. institutional occupancy and any other occupancy, except industrial occupancy,

C. one or more of,

1. business occupancy,

2. personal services occupancy, or

3. mercantile occupancy,

and any other occupancy, except assembly occupancy, institutional occupancy or industrial occupancy, where the building as constructed, enlarged or altered exceeds 600 square metres in gross area or three storeys,

D. residential occupancy that exceeds three storeys and any other occupancy, where the building as constructed, enlarged or altered exceeds 600 square metres in gross area, or

viii. any other occupancy where the building as constructed, enlarged or altered exceeds 600 square metres in gross area or three storeys, but a professional engineer may provide a design for the industrial occupancy of a mixed occupancy described in subparagraph vi.

4. An architect may perform or provide services that are within the practice of professional engineering in preparing or providing a design for and carrying out the general review of the construction enlargement or alteration of a building described in rule 2 or 3 where to do so does not constitute a substantial part of the services within the practice of professional engineering related to the construction, enlargement or alteration of the building and is necessary,

i. for the construction, enlargement or alteration of the building and is incidental to other her services provided as part of the practice of architecture by the architect in respect of the construction, enlargement or alteration of the building, or

ii. for coordination purposes.

5. A professional engineer may perform or provide services that are within the practice of architecture in preparing or providing a design for and carrying out the general review of the construction, enlargement or alteration of a building described in rule 1 or 3 where to do so does not constitute a substantial part of the services within the practice of architecture related to the construction, enlargement or alteration of the building and is necessary,

i. for the construction, enlargement or alteration of the building and is incidental to other services provided as part of the practice of professional engineering by the professional engineer in respect of the construction, enlargement or alteration of the building, or

ii. for coordination purposes.

6. Only an architect may carry out or provide the general review of the construction, enlargement or alteration of a building,

i. that is constructed, enlarged or altered in accordance with a design prepared or provided by an architect, or

ii. in relation to services that are provided by an architect in connection with the design in accordance with which the building is constructed, enlarged or altered.

7. Only a professional engineer may carry out or provide the general review of the construction, enlargement or alteration of a building,

i. that is constructed, enlarged or altered in accordance with a design prepared or provided by a professional engineer, or

ii. in relation to services that are provided by a professional engineer in connection with the design in accordance with which the building is constructed, enlarged or altered.

8. A professional engineer or an architect may act as prime consultant for the construction, enlargement or alteration of a building.

9. A reference in these rules to the provision of a design or services by a professional engineer applies equally to a holder of a certificate of authorization. R.S.O. 1990,c. P. 28, s. 12(6).

Idem
(7) Subsections (1) and (2) do not apply to prevent a person from carrying out a general review of the construction, enlargement or alteration of a building that does not or is not intended to take the place of a general review required to be done by a professional engineer. R.S.O.1990, c. P.28, s. 12(7).

Definitions

(8) In this section,

"assembly occupancy" means occupancy for gatherings of persons for civic, educational, political, recreational, religious, social, travel or other similar purpose, or for the consumption of food or drink; ("etablissement de reunion")

"building" means a structure consisting of a wall, roof and floor, or any one or more of them; ("batiment")

"building area" means the greatest horizontal area of a building within the outside surface of exterior walls or, where a firewall is to be constructed, within the outside surface of exterior walls and the centre line of firewalls; ("aire de batiment")

"business occupancy" means occupancy for the transaction of business; ("etablissement d'affaires")

"construction" means the doing of anything in the erection, installation, extension or repair of a building and includes the installation of a building unit fabricated or moved from elsewhere, and "constructed" has a corresponding meaning; ("construction", "construit")

"design" means a plan, sketch, drawing, graphic representation or specification intended to govern the construction, enlargement or alteration of a building or a part of a building; ("plan") "dwelling unit" means a room or suite of rooms used or intended to be used as a domicile by one or more persons and usually containing cooking, eating, living, sleeping and sanitary facilities; ("logement")

"fire separation" means a construction assembly that acts as a barrier against the spread of fire and that may or may not have a fire-resistance rating or a fire-protection rating; ("separation coupe-feu")

"firewall" means a type of fire separation of non-combustible construction that subdivides a building or separates adjoining buildings to resist the spread of fire and that has a fire-resistance rating as prescribed in the building code under the Building Code Act and has structural stability to remain intact under fire conditions for the fire-resistance time for which it is rated; ("mur coupe-feu")

"general review", in relation to the construction, enlargement or alteration of a building, means an examination of the building to determine whether the construction, enlargement or alteration is in general conformity with the design governing the construction, enlargement or alteration, and reporting thereon; ("examen de conformite")

"grade" means the lowest of the average levels finished ground adjoining each exterior wall of a building, but does not include localized depressions such as for vehicle or pedestrian entrances; ("niveau du sol")

"graphic representation" means a representation produced by electrical, electronic, photographic or printing methods and includes a representation produced on a video display terminal; ("representation graphiue")

"gross area" means the total area of all floors above grade measured between the outside surfaces of exterior walls or, where no access or building service penetrates a firewall, between the outside surfaces of exterior walls and the centre line of firewalls but in a residential occupancy where access or a building service penetrates a firewall, the measurement may be taken to the centre line of the firewall; ("surface hors-tout")

"industrial occupancy" means occupancy for assembling fabricating, manufacturing, processing, repairing or storing of goods or materials or for producing, converting, processing or storing of energy, waste or natural resources; ("etablissement industriel")

"institutional occupancy" means occupancy for the harbouring, housing or detention of persons who require special care or treatment on account of their age or mental or physical limitations or who are involuntarily detained; (establissement hospitalier, d'assistan 'eou de detention")

"mercantile occupancy" means occupancy or use for displaying or selling retail goods, wares or mercha- ndise ("etablissement commercial")

"personal services occupancy" means occupancy for the rendering or receiving of professional or personal services; ("etablissement de services personnels")

"residential occupancy" means occupancy for providing sleeping accommodation for persons but dos not include institutional occupancy, ("habitation") R.S.O. 1990, c. P.28, s. 12(8).

Proof of practice

(9) For the purposes of this section, proof of the performance of one act in the practice of

professional engineering on one occasion is sufficient to establish engaging in the practice of professional engineering. R.S.O. 1990 c. P 28 s. 12(9).

Corporation

13. A corporation that holds a certificate of authorization may provide services that are within the practice of professional engineering. RS.O. 1990, c. P. 28, s. 13.

Issuance of licence

14.(1) The Registrar shall issue a licence to a natural person who applies therefore in accordance with the regulation and,

(a) is a citizen of Canada or has the status of a permanent resident of Canada

(b) is not less than eighteen years of age;

(c) has complied with the academic requirements specified in the regulations for the issuance of the licence and has passed such examinations as the Council has set or approved in accordance with the regulations or is exempted therefrom by the Council;

(d) has complied with the experience requirements specified in the regulations for the issuance of the licence; and

(e) is of good character. R.S.O. 1990, c. P.28, s. 14(1).

Grounds for refusal to issue licence

(2) The Registrar may refuse to issue a licence to an applicant where the Registrar is of the opinion, upon reasonable and probable grounds, that the past conduct of the applicant affords grounds for belief that the applicant will not engage in the practice of professional engineering in accordance

with the law and with honesty and integrity.-R.S.O. 1990, c. P.28, s. 14(2).

Referral to committee

(3) The Registrar, on his or her own initiative, may refer and on the request of an applicant shall refer the application of the applicant for the issuance of a licence,

(a) to the Academic Requirements Committee for a determination as to whether or not the applicant has met the academic requirements prescribed by the regulations for the issuance of the licence;

(b) to the Experience Requirements Committee for a determination as to whether or not the applicant has met the experience requirements prescribed by the regulations for the issuance of the licence; or

(c) first to the Academic Requirements Committee and then to the Experience Requirements Committee for determinations under clauses (a) and (b). R.S.O. 1990, c. P.28, s. 14(3).

(4) REPEALED: 2001, c. 9, Sched. B, s. 11(18).

Hearing

(5) A committee shall receive written representations from an applicant but is not required to hold or to afford to any person a hearing or an opportunity to make oral submissions before making a determination under subsection (3). R.S.O. 1990, c. P.28, s. 14(5).

Notice of determination

(6) The Registrar shall give notice to the applicant of a determination by a committee under subsection (3) and, if the applicant is

rejected, the notice shall detail the specific requirements that the applicant must meet. R.S.O. 1990, c. P. 28, s. 14(6).

Provisional licence

(7) The Registrar shall issue a provisional licence, to be valid for one year, to a natural person who has applied for a licence in accordance with the regulations and has complied with all the requirements of subsection (1) except the Canadian experience requirement set out in paragraph 4 of section 33 of Regulation 941 of the Revised Regulations of Ontario, 1990. 2001, c. 9, Sche B, s. 11(19).

Issuance of certificate of authorization

15.(1) The Registrar shall issue a certificate of authorization to a natural person, a partnership or a corporation that applies therefore in accordance with the regulations of the requirements and qualifications for the issuance of the certificate of authorization set out in the regulations are met.

General and standard certificate

(2) Where the Registrar proposes to issue a certificate of authorization to an applicant, the Registrar shall issue a standard certificate of authorization or, where the primary function of the applicant is or will be to provide to the public services that are within the practice of professional engineering and the applicant requests a general certificate of authorization, the Registrar shall issue a general certificate of authorization to the applicant.

Partnership of corporations

(3) The Registrar shall issue a standard certificate of authorization to a partnership of corporations that applies therefore in

accordance with the regulations if at least one of the corporations holds a certificate of authorization.

Terms and conditions

(4) Where a holder of a temporary licence assumes responsibility for and supervises the practice of professional engineering related to the services provided by he holder of a certificate of authorization, the certificate of authorization is subject to the same terms and conditions prescribed by the regulations that apply to the temporary licence.

Suspension of effect of certificate of authorization

(5) A holder of a certificate of authorization ceases to be entitled to offer to the public or to provide to the public services that are within the practice of professional engineering as soon as there is no holder of a licence or a temporary licence who assumes responsibility for and supervises the practice of professional engineering provided by the holder of the certificate of authorization.

Notice to Registrar by holder of certifies of authorization

(6) The holder of a certificate of authorization must give notice to the Registrar when there ceases to be a holder of a licence or a temporary licence who assumes responsibility for and supervises the practice of professional engineering by the holder of the certificate of authorization and when the holder of the certificate of authorization designates another holder of a licence or a temporary licence to assume such responsibility and carry out such supervision.

Notice to Registrar by person in position of professional responsibility

(7) A holder of a licence or a temporary licence who ceases to be responsible for and to supervise the practice of professional engineering by a holder of a certificate of authorization as the person so designated by the holder of the certificate of authorization shall give notice of the cessation forthwith to the Registrar.

Past conduct

(8) The Registrar may refuse to issue or may suspend or revoke a certificate of authorization where the Registrar is of the opinion, upon reasonable and probable grounds,

(a) that the past conduct of a person who is in a position of authority or responsibility in the operation of the business of the applicant for or the holder of the certificate of authorization affords grounds for the belief that the applicant or holder will not engage in the business of providing services that are within the practice of professional engineering in accordance with the law and with honesty and integrity;

(b) that the holder of the certificate of authorization does not meet the requirements or the qualifications for the issuance of the certificate of authorization set out in the regulations; or

(c) that there has been a breach of a condition of the certificate of authorization. R.S.O. 1990, c. P.28, s. 15.

Issuance of licence or certificate of authorization on direction of Council

16. The Registrar shall issue a licence or a certificate of authorization upon a direction of the Council made in accordance with a recommendation by the Joint Practice Board. R.S.O. 1990, c. P. 28, s. 16.

Supervision by professional engineer

17.(1) It is a condition of every certificate of authorization that the holder of the certificate shall provide services that are within the practice of professional engineering only under the personal supervision and direction of a member of the Association or the holder of a temporary licence.

Professional responsibility of supervising professional engineer

(2) A member of the Association or a holder of a temporary licence who personally supervises and directs the providing of services within the practice of professional engineering by a holder of a ceertificate of authorization or who assumes responsibility for and supervises the practice of professional engineering related to the providing of services by a holder of a certificate of authorization is subject to the same standards of professional conduct and competence in respect of the services and the related practice of professional engineering as if the services were provided or the practice of professional engineering was engaged in by the member of the Association or the holder of the temporary licence. R.S.O. 1990, c. P.28, s. 17.

Issuance of temporary, provisional or limited licence

18.(1) The Registrar shall issue a temporary licence, provisional licence or a limited licence to a natural person who applies therefor in accordance with the regulations and who meets the requirements and qualifications for the issuance of the temporary licence, the provisional licence or the limited licence set out in the regulations, provided that, in the case of a limited or provisional licence, the applicant is a Canadian citizen or has the status of a permanent resident of Canada. 2001 c. 9, Sched.B, s. 11(20).

Grounds for refusal, suspension or revocation

(2) The Registrar may refuse to issue or may suspend or revoke a temporary licence, a provisional licenc or a limited licence where the Registrar is of the opinion, upon reasonable and probable grounds,

(a) that the past conduct of the applicant for or the holder of the temporary licence, the provisional licence or the limited licence affords grounds for the
belief that the applicant or holder will not engage in the practice of professional engineering in accordance with the law and with honesty and integrity;

(b) that the holder of the temporary licence, the provisional licence or the limited licence does not meet the requirements or the qualifications for the issuance of the temporary licence, the provisional licence or the limited licence set out in the regulations; or

(c) that there has been a breach of a condition of the temporary licence, the provisional licence or the limited licence. R.S.O. 1990, c. P.28, s. 18(2); 2001, c.9, Sched. B, s. 11(21).

Referral to committee

(3) Subsections 14 (3) to (6) (which relate to the Academic Requirements Committee and the Experience Requirements Committee) apply with necessary modifications in respect of an applicant for a temporary

licenc or a limited licence. R.S.O. 1990,c. P.28, s. 18(3).

Application of subs. (1)

(4) Subsection (1) does not apply in respect of a member of the Association or a holder of a certificate of authorization. R.S.O. 1990, c. P.28, s. 18(4).

Membership

(5) A holder of a temporary licence or a limited licence is not a member of the Association. R.S.O. 1990, c P. 28, s. 18(5).

Notice of proposal to revoke or refuse to renew

19.(1) Where the Registrar proposes,

(a) to refuse to issue a licence; or

(b) to refuse to issue, to suspend or to revoke a temporary licence, a provisional licence, a limited licence or a certificate of authorization, the Registrar shall serve notice of the proposal, together with written reasons therefore, on the applicant. R.S.O. 1990, c. P.28, s. 19(1); 2001, c. 9, Sched. B, s. 11(22).

Exception

(2) Subsection (1) does not apply in respect of a proposal to refuse to issue a licence, a temporary licence, a provisional licence or a limited licence where the applicant previously held a licence, a certificate of authorization, a temporary licence, a provisional licence or a limited licence that was suspended or revoked as a result of a decision of the Discipline Committee. 2001, c. 9, Sched. B.s. 11(23).

Notice

(3) A notice under subsection (1) shall state that the applicant is entitled to a hearing by

the Registration Committee if the applicant mails or delivers, within thirty days after the notice under subsection (1) is served on the applicant, notice in writing requiring a hearing by the Registration Committee and the applicant may so require such a hearing. R.S.O. 1990, c. P.28, s. 19(3).

Power of Registrar where no hearing

(4) Where the applicant does not require a hearing by the Registration Committee in accordance with subsection (3), the Registrar may carry out the proposal stated in the notice under subsection (1). R.S.O. 1990, c. P.28, s. 19(4).

Hearing by Registration Committee

(5) Where an applicant requires a hearing by the Registration Committee in accordance with subsection (3), the Registration Committee shall appoint a time for, give notice of and shall hold the hearing. R.S.O. 1990, c. P.28,s. 19(5).

Continuation on expiry of committee membership

(6) Where a proceeding is commenced before the Registration Committee and the term of office on the Council or on the committee of a member sitting for the hearing expires or is terminated other than for cause before the proceeding is disposed of but after evidence is heard, the member shall be deemed to remain a member of the Registration Committee for the purpose of completing the disposition of the proceeding in the same manner as if the member's term of office had not expired or been terminated. R.S.O. 1990, c. P.28, s. 19(6).

Powers of Registration Committee

(7) Following upon a hearing under this section in respect of a proposal by the

Registrar, the Registration Committee may, by order,

(a) where the committee is of the opinion upon reasonable grounds that the applicant meets the requirements and qualifications of this Act and the regulations and will engage in the practice of professional engineering or in the business of providing services that are within the practice of professional engineering with competence and integrity, direct Registrar to issue a licence, certificate of authorization, temporary licence, provisional licence or limited licence, as the case may be, to the applicant

(b) where the committee is of the opinion upon reasonable grounds that the applicant does not meet the requirements and qualifications of this Act and the regulations,

(i)   direct the Registrar to refuse to issue licence, certificate of authorization, temporary licence, provisional licence or limited licence, or to suspend or revoke the certificate of authorization issued to the applicant, as the case may be, or

(ii) where the committee is of the opinion upon reasonable grounds that the applicant will engage in the practice of professional engineering with competence and integrity, exempt the applicant from any of the requirements of this Act and the regulations and direct the Registrar to issue a licence, certificate of authorization, temporary licence, provisional licence or limited licence, as the case may be; or

© where the committee is of the opinion upon reasonable grounds that it is necessary in order to ensure that the applicant will engage in the practice of professional engineering or in the business of providing services that are within the practice of professional engineering with competence and integrity, direct the Registrar to issue a licence, certificate of authorization, temporary licence, provisional licence or limited licence, as the case may be, subject to such terms, conditions or limitations as the Registration Committee specifies. R.S.O. 1990, c. P.28, s. 19(7); 2001, c. 9, Sched. B, s. 11(24).

Extension of time for requiring hearing

(8) The Registration Committee may extend the time for the giving of notice requiring a hearing by an applicant under this section before or after the expiration of such time where it is satisfied that there are apparent grounds for granting relief to the applicant following upon a hearing and that there are reasonable grounds for applying for the extension, and the Registration Committee may give such directions as it considers proper consequent upon the extension. R.S.O., 1990, c. P.28, s. 19(8).

Parties

(9) The Registrar and the applicant who has required the hearing are parties to proceedings before the Registration Committee under this section. R.S.O. 1990 , c .P. 28, s. 19(9).

Opportunity to show compliance

(10) The applicant shall be given a reasonable opportunity to show or to achieve compliance before the hearing with all lawful requirements for the issue of the licence, the certificate of authorization, the temporary licence, the provisional licence or the limited licence. R.S.O. 1990, c. P.28, s. 19(10); 2001, c. 9, Sched. B, s. 11(25).

Examination of documentary evidence

(11) A party to proceedings under this section shall be afforded an opportunity to examine before the hearing any written or documentary evidence that will be produced or any report the contents of which will be given in evidence at the hearing. R.S.O. 1990, c. P.28, s. 19(11).

Members holding hearing not to have taken part in investigation, etc.

(12) Members of the Registration Committee holding a hearing shall not have taken part before the hearing in any investigation or consideration of the subject-matter of the hearing and shall not communicate directly or indirectly in relation to the subject-matter of the hearing with any person or with any party or representative of a party except upon notice to and opportunity for all parties to participate, but the Registration Committee may seek legal advice from an adviser independent from the parties and, in such case, the nature of the advice shall be made known to the parties in order that they may make submissions as to the law. R.S.O. 1990, c. P.28, s. 19(12).

Recording of evidence

(13) The oral evidence taken before the Registration Committee at a hearing shall be recorded and, if so required, copies of a transcript thereof shall be furnished upon the same terms as in the Superior Court of Justice. R.S.O.1990, c. P.28, s. 19(13); 2001, c. 9, Sched. B, s. 11(66).

Only members at hearing to participate in decision

(14) No member of the Registration Committee shall participate in a decision of the Registration Committee following upon a hearing unless he or she was present throughout the hearing and heard the evidence and argument of the parties. R.S.O. 1990, c. P.28, s. 19(14).

Release of documentary evidence

(15) Documents and things put in evidence at a hearing shall, upon the request of the person who produced them, be released to the person by the Registration Committee within a reasonable time after the matter in issue has been finally determined. R.S.O. 1990, c. P.28, s. 19(15).

Applicant

(16) In this section, "applicant" means applicant for a licence or applicant for or holder of a temporary licence, a provisional licence, a limited licence or a certificate of authorization. R.S.O. 1990, c. P.28, s. 19(16); 2001, c. 9, Sched. B, s. 1(26).

Fiduciary, etc., relationship between corporation and client

20. A corporation that holds a certificate of authorization has the same rights and is subject to the same obligations in respect of fiduciary, confidential and ethical relationships with each client of the corporation that exist at law between a member of the Association and his client. R.S.O. 1990, c. P.28, s. 20.

Registers

21.(1) The Registrar shall maintain one or more registers in which is entered every person who is licensed under this Act and every holder of a certificate of authorization, temporary licence, provisional licence or limited licence, identifying the terms, conditions and limitation attached to the licence, certificate of authorization, temporary licence, provisional licence or limited licence, and shall note on the register every revocation, suspension and cancellation or termination of a licence, certificate of authorization, temporary licence, provisional licence or

limited licence and such other information as the Registration Committee or Discipline Committee directs. R.S.O. 1990, c. P.28, s. 21(1); 2001, c. 9, Sched. B, s.1 1(27).

Inspection

(2) Any person has the right, during normal business hours, to inspect the registers maintained by the Registrar R.S.O. 1990, c. P.28, s. 21(2).

Copies

(3) The Registrar shall provide to any person, upon payment of a reasonable charge therefore, a copy of any part of the registers mentioned in subsection (1) maintained by the Registrar. R-S.O. 1990, c. R28, s. 21(3)

Cancellation for default of fees

22.(1) The Registrar may cancel a licence, certificate of authorization, temporary licence, provisional licence or limited licence for non-payment of any fee prescribed by the regulations or the by-laws after giving the member or the holder of the certificate of authorization, temporary licence, provisional licence or limited licence at least two months notice of the default and intention to cancel, subject to the continuing jurisdiction of the Association in respect of any disciplinary action arising out of the person's professional conduct while a member or holder.
R.S.O. 1990, c. P.28, s. 22(1); 2001, c. 9, S hed. B, s. 11(28).

Reinstatement

(2) A person who was a member of the Association or a holder of a certificate of authorization, temporary licence, provisional licence or limited licence whose

licence, certificate of authorization, temporary licence, provisional licence or limited licence was cancelled by the Registrar under subsection (1) is entitled to have the licence, certificate of authorization, temporary licence, provisional licence, limited licence reinstated upon compliance with requirements and qualifications prescribed by the regulations. R.S.O. 1990, c. P.28, s. 22(2); 2001, c. 9, Sched. B, s. 11(29).

Complaints Committee

23.(1) The Complaints Committee shall be composed of not fewer than three members of the Association appointed to the Committee by the Council, including at least one member of the Council who was appointed to the Council by the Lieutenant Governor in Council.

Idem

(2) No person who is a member of the Discipline Committee shall be a member of the Complaints Committee.

Chair

(3) The Council shall name one member of the Complaints Committee to be chair.

Quorum

(4) Three members of the Complaints Committee, of whom one shall be a person appointed to the Council by the Lieutenant Governor in Council, constitute a quorum. R.S.O. 1990, c. P.28, s. 23.

Duties of Complaints Committee

24.(1) The Complaints Committee shall consider and investigate complaints made by members of the public or members of the Association regarding the conduct or actions of a member of the Association or holder of a certificate of authorization, a temporary licence, a provisional licence or a limited licence, but no action shall be taken by the Committee under subsection (2) unless,

(a) a written complaint in a form that shall be provided by the Association has been filed with the Registrar and the member or holder whose conduct or actions are being investigated has been notified of the complaint and given at least two weeks in which to submit in writing to the Committee any explanations or representations the member or holder may wish to make concerning the matter; and

(b) the Committee has examined or has made every reasonable effort to examine all records and other documents relating to the complaint. R.S.O. 1990, c. P.28, s. 24(1); 2001, c. 9, Sched. B, s. 11(30).

Idem

(2) The Committee in accordance with the information it receives may,

(a) direct that the matter be referred, in whole or in part to the Discipline Committee;

(b) direct that the matter not be referred under clause (a); or

(c) take such action as it considers appropriate in the circumstances and that is not inconsistent with this Act or the regulations or by-laws. R.S.O. 1990, c. P. 28, s. 24(2).

Decision and reasons

(3) The Committee shall give its decision in writing to the Registrar for the purposes of subsection (4) and, where the decision is

made under clause (2)(b), its reasons therefore. R.S.O. 1990, c. P.28, s. 24(3)

Notice

(4)The Registrar shall send to the complainant and to the person complained against by prepaid first class mail a copy of the written decision made by the Complaints Committee and its reasons therefore, if any, together with notice advising the complainant of the right to apply to the Complaints Review Councillor under section 26. R.S.O. 1990, c. P.28, s. 24(4).

Hearing

(5) The Committee is not required to hold a hearing or to afford to any person an opportunity for a hearing or an opportunity to make oral submissions before making a decision or giving a direction under this section R.S.O. 1990,c. P28, s. 24(5).

Complaints Review Councillor

25.(1) There shall be a Complaints Review Councillor who shall be appointed by and from among the members of the Council appointed by the Lieutenant Governor in Council under clause 3(2)(c).

Idem

(2) The Complaints Review Councillor is not eligible to be a member of the Complaints Committee or the Fees Mediation Committee. R.S.O. 1990, c. P.28, s. 25

Powers of Complaints Review Councillor Examination by Complaints Review Councillor

26.(1) The Complaints Review Councillor may examine from time to time the

procedures for the treatment of complaints by the Association. R.S.O. 1990, c. P.28, .s. 26(1).

Review by Complaints Review Councillor

(2) Where a complaint respecting a member of the Association or a holder of a certificate of authorization, a temporary licence, a provisional licence or a limited licence has not been disposed of by the Complaints Committee within ninety days after the complaint is filed with the Registrar, upon application by the complainant or on his or her own initiative the Complaints Review Councillor may review the treatment of the complaint by the Complaints Committee. R.S.O. 1990, c. P. 28, s.26(2) 2001, c.9, Sched. B, s. 11(31).

Application to Complaints Review Councillor

(3) A complainant who is not satisfied with the handling by the Complaints Committee of a complaint to the Committee may apply to the Complaints Review Councillor for a review of the treatment of the complaint after the Committee has disposed of the complaint. R.S. O. 1990., c. P.28, s. 26(3).

No inquiry into merits

(4) In an examination or review in respect of the Association, the Complaints Review Councillor shall not inquire into the merits of any particular complaint made to the Association. R.S.O. 1990, c. P.28, s. 26(4).

Discretionary power of Complaints Review Councillor

(5) The Complaints Review Councillor in his or her discretion may decide in a particular case not to make a review or not to continue a review in respect of the Association where,

(a) the review is or would be in respect of the treatment of a complaint that was disposed of by the Association more than twelve months before the matter came to the attention of the Complaints Review Councillor, or

(b) in the opinion of the Complaints Review Councillor,

(i) the application to the Complaints Review Councillor is frivolous or vexatious or is not made in good faith, or

(ii) the person who has made application to the Complaints Review Councillor has not a sufficient personal interest in the subject-matter of the particular complaint. R.S.O. 1990, c. P.28, s. 26(5).

Notice

(6) Before commencing an examination or review in respect of the Association, the Complaints Review Councillor shall inform the Association of the intention to commence the examination or review. R.S.O. 1990, c. P.28, s. 26(6).

Office accommodation

(7) The Council shall provide to the Complaints Review Councillor such accommodation and support staff in the offices of the Association as are necessary to the performance of the powers and duties of the Complaints Review Councillor. R.S.O. 1990, c. P.28, s. 26(7).

Privacy

(8) Every examination or review by the Complaints Review Councillor in respect of the Association shall be conducted in private. R.S.O. 1990, c. P.28, s. 26(8).

Receipt of information

(9) In conducting an examination or review in respect of the Association, the Complaints Review Councillor may hear or obtain information from any person and may make such inquiries as he or she thinks fit. R.S.O. 1990, c. P.28. s. 26(9).

Hearing not required

(10) The Complaints Review Councillor is not required to hold or to afford to any person an opportunity for a hearing in relation to an examination, review or report in respect of the Association. R.S.O. 1990, c. P.28, s. 26(10).

Duty to furnish information

(11) Every person who is,

(a) a member of the Council;

(b) an officer of the Association;

(c) a member of a committee of the Association; or

(d) an employee of the Association, shall furnish to the Complaints Review Councillor such information regarding any proceedings or procedures of the Association in respect of the treatment of complaints made to the Association as the Complaints Review Councillor from time to time requires, and shall give the Complaints Review Councillor access to all records, reports, files and other papers and things belonging to or under the control of the Association or any of such persons and that relate to the treatment by the Association of complaints or any particular complaint. R.S.O. 1990, c. P. 28, s. 26(11).

Report by Complaints Review Councillor

(12) The Complaints Review Councillor shall make a report following upon each examination or review by him or her in respect of the Association. R.S.O. 1990, c. P. 28, s. 26(12).

Report following upon examination

(13) Where the report follows upon an examination of the procedure for the treatments by the Association, the Complaints Review Councillor shall transmit the report to the Council RS.O. 1990, c. P. 28, s. 26(13)

Report following upon review

(14) Where the report follows upon a review of the treatment of a complaint by the Association the Complaints Review Councillor shall transmit the report to the Council, to the complainant and to the person complained against. R.S.O. 1990, c. P.28, s. 26(14)

Report to Minister

(15) The Complaints Review Councillor may transmit a report following upon an examination or review to the Minister where, in the opinion of the Complaints Review Councillor, the report should be brought to the attention of the Minister. R.S.O. 1990, c. P.28, s. 26(15)

Recommendations

(16) The Complaints Review Councillor may include in a report following upon an examination or review his or her recommendations in respect of the procedure of the Association, either generally or with respect to the treatment of a particular complaint. R.S.O. 1990, c. P28. s.26(16).

Consideration by Council

(17) The Council shall consider each report, and any recommendations included in the report, transmitted to it by the Complaints Review Councillor and shall notify the Complaints Review Councillor of any action it has taken in consequence. R-S.O. 1990, c. P.28, s. 26(17).

Discipline Committee

27.(1) The Discipline Committee shall be composed of,
(a) at least one person appointed to the Discipline Committee by the Council from among the members of the Council elected to the Council;

(b) at least one person who is a member of the Association and who is a member of the Council appointed by the Lieutenant Governor in Council; and

(c) the persons appointed to the Committee by the Council from among the members of the Association who have not less than ten years experience in the practice of professional engineering. R.S.O.1990, c. P.28, s. 27(1).

Additional members

(1.1) The Discipline Committee may also include one or more persons appointed by the Council from among the members of the Council appointed by the Lieutenant Governor in Council under clause 3(2)(c). 2001, c. 9 Sched.B,s. 11(32).

Quorum and votes

(2) Five members of the Discipline Committee, of whom one shall be a person appointed to the Council by the Lieutenant Governor in Council and one shall be a person elected to the Council, constitute a

quorum, and all disciplinary decisions require the vote of a majority of the members of the Discipline Committee present at the meeting. R.S.O. 1990, c. P.28, s. 27(2).

Disability of member

(3) Where the Discipline Committee commences a hearing and the member or members thereof who are appointed to the Council by the Lieutenant Governor in Council or who are elected members of the Council become unable to continue to act, the remaining members may complete the hearing despite the absence of the member or members and may render a decision as effectually as if all members of the Discipline Committee were present throughout the hearing, despite the absence of a quorum of the Committee. R.S.O. 1990, c. P.28, s. 27(3).

Chair

(4) The members of the Discipline Committee shall name one of them to be the chair of the Discipline Committee. R.S.O. 1990, c. P.28, s. 27(4).

Reference by Council or Executive Committee

(5) The Council or the Executive Committee, by resolution, may direct the Discipline Committee to hold a hearing and determine any allegation of professional misconduct or incompetence on the part of a member of the Association or a holder of a certificate of authorization, a temporary licence, a provisional licence or a limited licence specified in the resolution. R.S.O. 1990, P.28, s. 27(5); 2001, c. 9, Sched. B, s. 11(33)

Chair may refer matter to panel

(6) When a matter is referred to the Discipline Committee for hearing and determination, the chair may,

(a) select from among the members of the Committee a panel composed of at least one person described in clause (1)(a), at least one person described in clause (1)(b), at least one person described in clause (1)(c), and, if the Council has made an appointment under subsection (1.1), at least one described in that subsection;

(b) designate one of the members of the panel to chair it;

(c) refer the matter to the panel for hearing and determination; and

(d) set a date, time and place for the hearing., 2001, c. 9, Sched. B ,s. 11(34).

Powers of panel

(7) A panel established under subsection (6) has all the powers and responsibilities of the Discipline Committee with respect to the hearing and determination of the matter referred to the panel 2001, c. 9, Sched. B, s. 11(34).

Duties and powers of Discipline Committee
Duties of Discipline Committee

28.(1) The Discipline Committee shall,

(a) when so directed by the Council, the Executive Committee or the Complaints Committee hear and determine allegations of professional misconduct or incompetence against a member of the Association or a holder of a certificate of authorization a temporary licence, a provisional licence or a limited licence;

(b) hear and determine matters referred to it under section 24, 27 or 37; and

(c) perform such other duties as are assigned to it by the Council. R.S.O. 1990, c. P.28,s. 28(1) 2001, c. 9, Sched. B, s. 11(35).

Professional misconduct

(2) A member of the Association or a holder of a certificate of authorization, a temporary licence, a provisional licence or a limited licence may be found guilty of professional misconduct by the Committee if,

(a) the member or holder has been found guilty of an offence relevant to suitability to practise, upon proof of such conviction;

(b) the member or holder has been guilty in the opinion of the Discipline Committee of professional misconduct as defined in the regulations. R.S.O. 1990, c. P.28, s. 28(2); 2001, c. 9, Sched. B, s. 11(36)

Incompetence

(3) The Discipline Committee may find a member of the Association or a holder of a temporary licence, a provisional licence or a limited licence to be incompetent if in its opinion,

(a) the member or holder has displayed in his or her professional responsibilities a lack of knowledge, skill or judgment or disregard for the welfare of the public of a nature or to an extent that demonstrates the member or holder is unfit to carry out the responsibilities of a professional engineer; or

(b) the member or holder is suffering from a physical or mental condition or disorder of a nature and extent making it desirable in the interests of the public or the member or holder that the member or holder no longer be permitted to engage in the practice of professional engineering or that his or her practice of professional engineering be restricted. R.S.O. 1990, c. P.28, s. 28(3); 2001, c. 9, Sched.B.s. 11(37).

Powers of Discipline Committee

(4) Where the Discipline Committee finds a member of the Association or a holder of a certificate of authorization, a temporary licence, a provisional licence or a limited licence guilty of professional misconduct or to be incompetent it may, by order,

(a) revoke the licence of the member or the certificate of authorization, temporary licence, provisional licence or limited licence of the holder;

(b) suspend the licence of the member or the certificate of authorization, temporary licence, provisional licence or limited licence of the holder for a stated period, not exceeding 24 months;

(c) accept the undertaking of the member or holder to limit the professional work of the member or holder in the practice of professional engineering to the extent specified in the undertaking;

(d) impose terms, conditions or limitations on the licence or certificate of authorization, temporary licence, provisional licence or limited licence, of the member or holder, including but not limited to the successful completion of a particular course or courses of study, as are specified by the Discipline Committee;

(e) impose specific restrictions on the licence or certificate of authorization, temporary licence, provisional licence or limited licence, including but not limited to,

(i) requiring the member or the holder of the certificate of authorization, temporary licence, provisional licence or limited licence to engage in the practice of professional engineering only under the personal supervision and direction of a member,

(ii) requiring the member to not alone engage in the practice of professional engineering,

(iii) requiring the member or the holder of the certificate of authorization, temporary licence, provisional licence or limited licence to accept periodic inspections by the Committee or its
delegate of documents and records in the possession or under the control of the member or the holder in connection with the practice of professional engineering,

(iv) requiring the member or the holder of the certificate of authorization, temporary licence, provisional licence or limited licence to report to the Registrar or to such committee of the Council as the Discipline Committee may specify on such matters in respect of the member's or holder's practice for such period of time, at such times and in such form, as the Discipline Committee may specify;

(f) require that the member or the holder of the certificate of authorization, temporary licence, provisional licence or limited licence be reprimanded, admonished or counselled and, if considered warranted, direct that the fact of the reprimand, admonishment or counselling be recorded on the register for a stated or unlimited period of time;

(g) revoke or suspend for a stated period of time the designation of the member or holder by the Association as a specialist, consulting engineer or otherwise;

(h) impose such fine as the Discipline Committee considers appropriate, to a maximum of $5,000, to be paid by the member of the Association or the holder of

the certificate of authorization, temporary licence, provisional licence or limited licence to the Treasurer of Ontario for payment into the consolidated Revenue Fund;

(i) subject to subsection (5) in respect of orders of revocation or suspension, direct that the finding and the order of the Discipline Committee be published in detail or in summary and either with or without including the name of the member or holder in the official publication of the Association and in such other manner or medium as the Discipline Committee considers appropriate in the particular case;

(j) fix and impose costs to be paid by the member or the holder to the Association

(k) direct that the imposition of a penalty be suspended or postponed for such period and upon such terms or for such purpose as the Discipline Committee
may specify, including but not limited to,

(i) the successful completion by the member or the holder of the temporary licence, provisional licence or limited licence of a particular course or courses of study,

(ii) the production to the Discipline Committee of evidence satisfactory to it that any physical or mental handicap in respect of which the penalty was imposed has been overcome, or any combination of them. 2001, c. 9, Sched. B, s. 11(38).

Publication of revocation or suspension

(5) The Discipline Committee shall cause an order of the Committee revoking or suspending a licence or certificate of authorization, temporary licence, provisional licence or limited licence to be published, with or without the reasons therefore, in the official publication of the

Association together with the name of the member or holder of the revoked or suspended licence or certificate of authorization, temporary licence, provisional licence or limited licence. R.S.O. 1990, c. P.28, s. 28(5); 2001, c. 9, Sched. B, s. 11(39).

Publication on request

(6) The Discipline Committee shall cause a determination by the Committee that an allegation of professional misconduct or incompetence was unfounded to be published in the official publication of the Association, upon the request of the member of the Association or the holder of the certificate of authorization, temporary licence, provisional licence or limited licence against whom the allegation was made. R.S.O. 1990, c. P. 28, s. 28(6); 2001, c. 9, Sched. B, s. 11(40).

Costs

(7) Where die Discipline Committee is of the opinion that the commencement of the proceedings was unwarranted, the Committee may order that the Association reimburse the member of the Association or the holder of the certificate of authorization, temporary licence, provisional licence or limited licence for the persons costs or such portion thereof as the Discipline Committee fixes. R.S.O. 1990, c. P. 28, s. 28(7); 2001, c. 9, Sched. B, s. 11(41).

Stay of decision on appeal

29.(1) Where the Discipline Committee revokes, suspends or restricts a licence, temporary licence, provisional licence or limited licence on the grounds of incompetence the decision takes effect immediately even if an appeal is taken from the decision, unless the court to which the appeal is taken otherwise orders, and, where

the court is satisfied that it is appropriate in the circumstances, the court may so order. R.S.O. 1990, c. P.28, s. 29(1); 2001, c. 9, Sched. B, s. 11(42).

Stay of decision on appeal, professional misconduct

(2) Where the Discipline Committee revokes, suspends or restricts a licence or a certificate of authorization, temporary licence, provisional licence or limited licence on grounds other than for incompetence, the order does not take effect until the time for appeal from the order has expired without an appeal being taken or, if taken, the appeal has been disposed of or abandoned, unless the Discipline Committee otherwise orders, and, where the Committee considers that it is appropriate for the protection of the public, the Committee may so order. R.S.O.1990, c. P. 28, s. 29(2); 2001, c. 9, Sched. B, s. 11(43).

Discipline proceedings

30.(1) In proceedings before the Discipline Committee, the Association and the member of the Association or the holder of a certificate of authorization, a temporary licence, a provisional licence or a limited licence whose conduct is being investigated in the proceedings are parties to the proceedings. R.S.O. 1990, c. P.28, s. 30(1), 2001, c. 9, Sched. B, s. 11(44).

Examination of documentary evidence

(2) A member or holder of a certificate of authorization, temporary licence, a provisional licence or a limited licence whose conduct is being investigated in proceedings before the Discipline Committee shall be afforded an opportunity to examine before the hearing any written or documentary evidence that will be produced or any report the contents of which will be

given in evidence at the hearing. R.S.O. 1990, c. P.28, s. 30(2), 2001,c. 9, Sched. B, s. 11(45).

Members holding hearing not to have taken part in investigation, etc.

(3) Members of the Discipline Committee holding a hearing shall not have taken part before the hearing in any investigation of the subject-matter of the hearing other than as a member of the Council considering the referral of the matter to the Discipline Committee or at a previous hearing of the Committee, and shall not communicate directly or indirectly in relation to the subject matter of the hearing with any person or with any party or representative of a party except upon notice to and opportunity for all parties to participate, but the Committee may seek legal advice from an adviser independent from the parties and, in such case, the nature of the advice shall be made known to the parties in order that they may make submissions as to the law. R.S 0. 1990, c. P.28, s. 30(3).

Public hearings

(4) Hearings of the Discipline Committee shall be open to the public, subject to subsection (4.1). 2001, c 9, Sched. B, s.1 1(46).

Exception

(4.1) The Discipline Committee may order that the public be excluded from all or part of a hearing if the following conditions are satisfied:

1. The person whose conduct is being investigated delivers to the Registrar, before the day fixed for the hearing or part, a written request that the hearing or part be closed.

2. The Discipline Committee is satisfied that,

i. matters involving public security may be disclosed at the hearing or part, or

ii. financial or personal or other matters may be disclosed at the hearing or part, of such a nature that the desirability of avoiding public disclosure of these matters in the interest of any person affected or in the public interest outweighs the desirability of adhering to the principle that hearings be open to the public. 2001 ,c.9, Sched .B ,s. 11(46).

Recording of evidence

(5) The oral evidence taken before the Discipline Committee shall be recorded and, if so required, copies of a transcript thereof shall be furnished only to the parties upon the same terms as in the Superior Court of Justice. R.S.O. 1990, c. P.28, s. 30(5); 2001, c. 9, Sched. B, s. 11(66).

Evidence

(6) Despite the Statutory Powers Procedure Act, nothing is admissible in evidence before the Discipline Committee that would be inadmissible in a court in a civil case and the findings of the Discipline Committee shall be based exclusively on evidence admitted before it. R.S.O. 1990, c. P.28, s. 30(6).

Only members at hearing to participate in decision
(7) No member of the Discipline Committee shall participate in a decision of the Committee following upon a hearing unless he or she was present throughout the hearing and heard the evidence and argument of the parties. R.S.O. 1990, c. P.28, s. 30(7).

Release of documentary evidence

(8) Documents and things put in evidence at a hearing of the Discipline Committee shall, upon the request of the party who produced them, be returned by the Committee within a reasonable time after the matter in issue has been finally determined. R.S.O. 1990, c. P.28, s. 30(8).

Continuation on expiry of Committee membership

(9) Where a proceeding is commenced before the Discipline Committee and the term of office on the Council or on the Committee of a member sitting for the hearing expires or is terminated, other than for cause, before the proceeding is disposed of but after evidence has been heard, the member shall be deemed to remain a member of the Discipline Committee for the purpose of completing the disposition of the proceeding in the same manner as if the term of office had not expired or been terminated. R.S.O. 1990, c. P.28, s. 30(9).

Service of decision of Discipline Committee

(10) Where the Discipline Committee finds a member of the Association or a holder of a certificate of authorization, temporary licence, provisional licence, or limited licence guilty of professional misconduct or incompetence, a copy of the decision shall be served upon the person complaining in respect of the conduct or action of the member or holder. R.S.O. 1990, c. P. 28, s. 30(10); 2001,c. 9, Sched. B, s. 11(47).

Appeal to court

31-(1) A party to proceedings before the Registration Committee or the Discipline Committee may appeal to the Divisional Court, in accordance with the rules of court, from the decision or order of the committee.

Certified copy of record

(2) Upon the request of a party desiring to appeal to the Divisional Court and upon payment of the fee therefore the Registrar shall furnish the party with a certified copy of the record of the proceedings, including the documents received in evidence and the decision or order appealed from.

Powers of court on appeal

(3) An appeal under this section may be made on questions of law or fact or both and the court may affirm or may rescind the decision of the committee appealed from and may exercise all powers of the committee and may direct the committee to take any action which the committee may take and as the court considers proper, and for such purposes the court may substitute its opinion for that of the committee or the court may refer the matter back to the committee for rehearing, in whole or in part, in accordance with such directions as the court considers proper. R.S.O. 1990, c. P.28, s. 31.

Fees Mediation Committee

32.(1) No person who is a member of the Complaint committee or the Discipline Committee shall be a member of the Fees Mediation Committee. R.S.O.1990. c. P. 28, s. 32(1).

Duties of Fees Mediation Committee

(2) The Fees Mediation Committee,

(a) shall, unless the Committee considers it inappropriate to do so, mediate any written complaint by a client of a member of the Association or of a holder of a certificate of authorization, a temporary licence, a provisional licence or a limited licence in respect of a fee charged for professional engineering services provided to the client; and

(b) shall perform such other duties as are assigned to it by the Council. R.S.O. 1990, c. P. 28, s. 32(2); 2001, c.9, Sched. B, s. 11(48).

Arbitration by Fees Mediation Committee

(3) The Fees Mediation Committee, with the written consent of all parties to the dispute, may arbitrate a dispute in respect of a fee between a client and a member of the Association or a holder of a certificate of authorization, temporary licence, provisional licence or limited licence and in that case the decision of the Fees Mediation Committee is final and binding on all parties to the dispute. R.S.O, 1990, c. P.28, s. 32(3); 2001, c. Sched. B.s. 11(49).

Procedure

(4) Where the Fees Mediation Committee acts as arbitrator under subsection (3), the Arbitrations Act does not apply. R.S.O. 1990, c. P.28, s. 32(4)

Enforcement
(5) A decision by the Fees Mediation Committee under subsection (3), exclusive of the reasons therefore, certified by the Registrar, may be filed with the Superior Court of Justice and when filed the decision may be enforced in the same manner as a judgment of the court. R.S.O. 1990, c. P.28, s. 32(5); 2001, c. 9, Sched. B, s. 11(66).

Registrar's investigation

33.(1) Where the Registrar believes on reasonable and probable grounds that a member of the Association or a holder of a certificate of authorization, a temporary licence, provisional licence or limited licence has committed an act of professional misconduct or incompetence or that there is cause to refuse to issue or to suspend or revoke a certificate of authorization, the Registrar by order may appoint one or more persons to investigate whether such act has occurred or there is such cause, and the person or persons appointed shall report the result of the investigation to the Registrar. R.S.O. 1990, c. P.28, s. 33(1);200 I.e. 9, Sched. B, s. 11(50).

Powers of investigator

(2) For purposes relevant to the subject-matter of an investigation under this section, the person appointed to make the investigation may inquire into and examine the practice of the member or holder of the certificate of authorization, temporary licence, provisional licence or limited licence in respect of whom the investigation is being made and may, upon production of his or her appointment, enter at any reasonable time the business premises of the member or holder and examine books, records, documents and things relevant to the subject-matter of the investigation and, for the purposes of the inquiry, the person making the investigation has the powers of a commission under Part II of the Public Inquiries Act, which Part applies to such inquiry as if it were an inquiry under that Act. R.S.O. 1990, c. P.28, s. 33(2); 2001, c. 9, Sched. B, s. 11(51).

Obstruction of investigator

(3) No person shall obstruct a person appointed to make an investigation under this section or withhold from him or her or conceal or destroy any books, records, documents or things relevant to the subject-matter of the investigation. R.S.O. 1990, c. P.28, s. 33(3).

Order by provincial judge

(4) Where a provincial judge is satisfied on evidence upon oath,

(a) that the Registrar had grounds for appointing and by order has appointed one or more persons to make an investigation; and

(b) that there is reasonable ground for believing there are in any building, dwelling, receptacle or place any books, records, documents or things relating to the member of the Association or holder of a certificate of authorization, a temporary licence, a provisional licence or a limited licence whose affairs are being investigated and to the subject-matter of .investigation, the provincial judge may issue an order authorizing the person or persons making the investigation, together with such police officer or officers as they call upon to assist them, to enter and search, by force if necessary, such building, dwelling , receptacle or place for such books, records, documents or things and to examine them. R.S.O. 1990, c.P.28, s. 33(4); 2001, c. 9, Sched. B, s. 11(52).

Execution of order

(5) An order issued under subsection (4) shall be executed at reasonable times as specified in the order. R.S.O.1990, c. P.28, s. 33(5).

Expiry of order

(6) An order issued under subsection (4) shall state the date on which it expires, which shall be a date not later than fifteen days after the order is issued. R.S.O. c. P.28, s. 33(6).

Application without notice

(7) A provincial judge may receive and consider an application for an order under subsection (4) without notice to and in the absence of the member of the Association or holder of a certificate of authorization, temporary licence, provisional licence or limited licence whose affairs are being investigated. R.S.O. 1990, c. P.28, s. 33(7); 2001, c. 9, Sched. B, s. 11(53).

Removal of books, etc.

(8) Any person making an investigation under this section may, upon giving a receipt therefor, remove any books, records, documents or things examined under this section relating to the member or holder whose practice is being investigated and to the subject-matter of the investigation for the purpose of making copies of such books, records or documents, but such copying shall be carried out with reasonable dispatch and the books, records or documents in question shall be promptly thereafter returned to the member or holder whose practice is being investigated. R.S.O. 1990, c. P.28 s. 33(8).

Admissibility of copies

(9) Any copy made as provided in subsection (8) and certified to be a true copy by the person making the investigation is admissible in evidence in any action, proceeding or prosecution as proof, in the absence of evidence to the contrary, of the original book, record or document and its contents. R.S.O. 1990, c. P.28, s. P. 3(9).

Report of Registrar

(10) The Registrar shall report the results of the investigation to the Council or such committee as the Registrar considers appropriate. R.S.O. 1990, c. P28, s. 33 (10)

Liability insurance

34. It is a condition of every certificate of authorization that the holder of the certificate shall not offer or provide to

the public services that are within the practice of professional engineering unless the holder is insured in respect of professional liability in accordance with the regulations. R.S.O. 1990, c. P.28, s. 34.

Insurance claims

35.(1) In this section, "insurer" means a person offering insurance in respect of liability incurred in the practice of professional engineering.

Information re insurance claims

(2) Upon the request of the Registrar, an insurer shall furnish to the Registrar all documents that relate to a claim for indemnity in respect of the practice of professional engineering and that are in the possession or under the control of the insurer and have been prepared by a professional engineer and relate to engineering matters.

Exception

(3) Subsection (2) does not apply in respect of a document prepared by an insured person related to a claim for indemnity in respect of the practice of professional engineering by the insured person.

Transmittal of information

(4) The Registrar may forward any information referred to in subsection (2) to the Council or to such committee as the Registrar considers appropriate. R.S.O. 1990, c. P.28, s. 35.

Surrender of revoked licence or certificate

36. Where a licence, certificate of authorization, temporary licence, provisional licence or limited licence is revoked or cancelled, the former holder thereof shall forthwith deliver the licence, certificate of authorization, temporary licence, provisional licence or limited licence and related seal to the Registrar. R.S.O. 1990, c. P.28, s. 36; 2001,c.9, Sched. B, s. 11(54).

Application after revocation or suspension

Application for licence, etc., after revocation

37.(1) A person whose licence, certificate of authorization, temporary licence, provisional licence or limited licence has been revoked for cause under this Act, or whose membership has been cancelled for cause under a predecessor of this Act, may apply in writing to the Registrar for the issuance of a licence, certificate of authorization, temporary licence, provisional licence or limited licence, but such application shall not be made sooner than two years after the revocation. R.S.O. 1990, c. P. 28, s. 37(1); 2001, c .9, Sched .B, s. 11(55).

Removal of suspension

(2) A person whose licence, certificate of authorization, temporary licence, provisional licence or limited licence has been suspended for cause under this Act, or whose membership has been suspended for cause under a predecessor of this Act, may apply in writing to the Registrar for the removal of the suspension, but, where the suspension is for more than one year, the application shall not be made sooner than one year after the commencement of the suspension. R.S.O. 1990, c. P. 28, s. 37(2); 2001, c. 9, Sched. B, s. 11(56).

Reference to Discipline Committee

(3) The Registrar shall refer an application under subsection (1) or (2) in respect of a

licence or a certificate of authorization, a temporary licence, a provisional licence or a limited licence to the Discipline Committee which shall. hold a hearing respecting and decide upon the application, and shall report its decision and reasons to the Council and the applicant. R.S.O. 1990, c. P.28, s. 37(3); 2001, c. 9, Schcd. B, s. 11(57).

Procedures

(4) The provisions of this Act applying to hearings by the Registration Committee, except section 31, apply with necessary modifications to proceedings of the Discipline Committee or the Registration Committee under this section. R.S.O. 1990, c. P.28, s. 37(4).

Confidentiality

38.(1) Every person engaged in the administration of this Act, including any person making an examination or review under section 26 or an investigation under section 33, shall preserve secrecy with respect to all matters that come to his or her knowledge in the course of his or her duties, employment, examination, review or investigation and shall not communicate any such matters to any other person except,
(a) as may be required in connection with the administration of,

(i) this Act and the regulations and by- laws, or

(ii) the Architects Act, and the regulations and by-laws under that Act, or any proceedings under

(iii) this Act or the regulations, or

(iv) the Architects Act, or the regulations under that Act;

(b) to his or her counsel; or

(c) with the consent of the person to whom the information relates. R.S.O. 1990, c. P.28, s. 38(1)

Testimony in civil action

(2) No person to whom subsection (1) applies shall be required to give testimony or to produce any book, record, document or thing in any action or proceeding with regard to information obtained in the course of his or her duties, employment, examination, review or investigation except in a proceeding under this Act or the regulations or by-laws or a proceeding under the Architects Act or the regulations or by-laws under that Act. R.S.O. 1990, c. P.28, s. 38(2).

Offence, penalty

(3) Every person who contravenes subsection (1) is guilty of an offence and on conviction is liable to a fine of not more than $10,000. 2001, c. 9, Sched. B, s. 11(58).

Limitation

(4) No proceeding shall be commenced in respect of an offence under subsection (1) after the expiration of two years after the date on which the offence was, or is alleged to have been, committed. 2001, c. 9, Sched. B, s. 11(58).

Order directing compliance

39-(1) Where it appears to the Association that any person does not comply with this Act or the regulations, despite the imposition of any penalty in respect of such non-compliance and in addition to any other rights it may have, the Association may apply to a judge of the Superior Court of Justice for an order directing the person to comply with the provision, and upon the application the judge may make the order or such other order as the judge thinks fit. R.S.O. 1990, c.P.28,s. 39(1); 2001,c. 9, Sched. B, s. 11(66).

Appeal

(2) An appeal lies to the Court of Appeal from an order made under subsection (1). R.S.O. 1990, c. P28, s. 39(2).

Penalties

40.(1) Every person who contravenes section 12 is guilty of an offence and on conviction is liable for the first offence to a fine of not more than $25,000 and for each subsequent offence to a fine of not more than $50,000. R.S.O. 1990, c. P. 28, s. 40(1).

Idem, use of term "professional engineer", etc.

(2) Every person who is not a holder of a licence or a temporary licence and who,

(a) uses the title "professional engineer" or "ingenieur" or an abbreviation or variation thereof as an occupational or business designation;

(a.l)uses the title "engineer" or an abbreviation of that title in a manner that will lead to the belief that the person may engage in the practice of professional engineering;

(b) uses a term, title or description that will lead to the belief that the person may engage in the practice of professional engineering; or

(c) uses a seal that will lead to the belief that the person is a professional engineer, is guilty of an offence and on conviction is liable for the first offence to a fine of not

more than $10,000 and for each subsequent offence to a fine of not more than $25,000. R.S.O. 1990, c. P.28, s. 40(2); 2001, c. 9, Sched. B, s. 11(59).

Onus of proof

(2.1)In a proceeding for an alleged contravention of clause (2)(a. l), the burden of proving that the use of the title or abbreviation will not lead to the belief referred to is on the defendant, unless the defendant's use of the title or abbreviation is authorized or required by an Act or regulation. 2001, c. 9 Sched, B, s. 11(60).

Idem, services of professional engineer

(3) Every person who is not acting under and in accordance with a certificate of authorization and who,
(a) uses a term, title or description that will lead to the belief that the person may provide to the public services that are within the practice of professional engineering; or

(b) uses a seal that will lead to the belief that the person may provide to the public services that are within the practice of professional engineering, is guilty of an offence and on conviction is liable for the first offence to a fine of not more than $10,000 and for each subsequent offence to a fine of not more than $25,000. R.S.O. 1990, c. P. 28, s. 40(3).

Idem

(4)Any person who obstructs a person appointed to make an investigation under section 33 in the course of his or her duties is guilty of an offence and on conviction is liable to a fine of not more than $10,000. R.S.O. 1990, c. P.28, s. 40(4).

Idem, director or officer of corporation

(5) Where a corporation is guilty of an offence under subsection (1), (2), (3) or (4),

every director or officer of the corporation who authorizes, permits or acquiesces in the offence is guilty of an offence and on conviction is liable to a fine of not more than $50,000. R.S.O. 1990, c. P.28, s. 40(5).

Idem, partner

(6) Where a person who is guilty of an offence under subsection (1), (2), (3) or (4) is a member or an employee of a partnership, every member of the partnership who authorizes, permits or acquiesces in the offence is guilty of an offence and on conviction is liable to a fine of not more than $50,000. R.S.0. 1990. c. P.28, s.40 (6).
1990,

Limitation

(7) Proceedings shall not be commenced in respect of an offence under subsection (1), (2), (3), (4), (5) or (6) after two years after the date on which the offence was or is alleged to have been, committed. R.S.O 1990, c. P.28, s. 40(7).

Application of subs. (2)

(8) Subsection (2) does not apply to a holder of a limited licence who uses a term, title or description: authorized or permitted by the regulations. R-S.O. 199 c. P. 28, s. 40(8).

Offences involving falsity

Falsification of documents
41.(1) Any person who makes or causes to be made a willful falsification in a matter relating to a register or issues a false licence, certificate, temporary licence, provisional licence, limited licence or document with respect to registration is guilty of an offence and on conviction is liable to a fine of not more than $10,000.

R.S.O. 1990, c. P.28, s. 41(1); 2001, c. 9, Sched. B, s. 11(61).

Offences for false representation

(2) Every person who willfully procures or attempts to procure the issuance of a licence, a certificate of authorization, a temporary licence, a provisional licence or a limited licence under this Act by knowingly making a false representation or declaration or by making a fraudulent representation or declaration, either orally or in writing, is guilty of an offence and on conviction is liable to a fine of not more than $10,000, and every person knowingly aiding and assisting such person therein is guilty of an offence and on conviction is liable to a fine of not more than $10,000. R.S.O. 1990, c. P. 28, s. 41(2); 2001, c. 9, Sched. B,s. 11(62).

Limitation

(3) No proceeding shall be commenced in respect of an offence under subsection (1) or (2) after the expiration of two years after the date on which the offence was, or is alleged to have been, committed. 2001, c. 9, Sched. B, s. 11(63).

Onus of proof respecting licensing

42. Where licensing or the holding of a certificate of authorization, a temporary licence, a provisional licence or a limited licence or acting under and in accordance with a certificate of authorization under this Act is required to permit the lawful doing of an act or thing, if in any prosecution it is proven that the defendant has done the act or thing, the burden of proving that the defendant was so licensed or that the defendant held a subsisting certificate of authorization, temporary licence, provisional licence or limited licence or that the defendant acted under and in accordance with a certificate of authorization under this

Act rests upon the defendant. R.S.O. 1990, c. P.28, s. 42; 2001, c. 9, Sched. B, s. 11(64).

Service of notice or document

43.(1) A notice or document under this Act or the regulations is sufficiently given, served or delivered if delivered personally or by mail.

Idem

(2)Where a notice or document under this Act or the regulations is sent to a person by mail addressed to the person at the last address of the person in the records of the Association, there is a rebuttable presumption that the notice or document is delivered to the person on the tenth day after the day of mailing. R.S.O. 1990, c. P. 28, s. 43.

Registrar's certificate as evidence

44. Any statement containing information from the records required to be kept by the Registrar under this Act, purporting to be certified by the Registrar under the seal of the Association, is admissible in evidence in all courts as proof, in the absence of evidence to the contrary, of the facts stated therein without proof of the appointment or signature of the Registrar and without proof of the seal. R.S.O. 1990, c. P.28, s. 44.

Immunity and indemnity
Immunity of Association

45.(1) No action or other proceeding for damages shall be instituted against the Association, a committee of the Association, or a member of the Association or committee of the Association, or an officer, employee, agent or appointee of the Association for any act done in good faith in the performance or

intended performance of a duty or in the exercise or intended exercise of a power under this Act, a regulation or a by-law, or for any neglect or default in the performance or exercise in good faith of such duty or power.

Councillor indemnified in suits respecting execution of office

(2) Every member of the Council, a committee of the Association and every officer and employee of the Association, and the person's heirs, executors and administrators, , and estate and effects, respectively, may, with the consent of the Association, given by the members of the Association, from time to time and at all times be indemnified and saved harmless out of the funds of the Association, from and against,

(a) all costs, charges and expenses whatsoever that the person sustains or incurs in or about any action, suit or proceeding that is brought, commenced or prosecuted against the person for or in respect of any act, deed, matter or thing whatsoever, made, done or permitted by the person, in or about the execution of the duties of the person's office; and

(b) all other costs, charges and expenses that the person sustains or incurs in or about or in relation to the affairs thereof, except such costs, charges or expenses as are occasioned by the person's own willful neglect or default. R.S.O. 1990, c. P.28, s. 45.

Limitation of action
46. REPEALED: 2002, c. 24, Sched. B, s. 25. See 2002, c.24, Sched. B, ss. 25, 51.

Joint Practice Board

47.(1) The Council shall appoint to the Joint Practice Board (composed of the chair, three members representing the Ontario Association of Architects and three members representing the Association of Professional Engineers of Ontario) the three members of the Joint Practice Board representing the Association and shall prescribe the term of each appointment.

Recommendation

(2) The Joint Practice Board may recommend to the Council that the Council direct the Registrar to issue a licence or a certificate of authorization to a holder of a certificate of practice issued under the Architects Act.

Direction by Council

(3) The Council, upon the recommendation of the Joint Practice Board, may direct the Registrar to issue a licence or a certificate of authorization to a holder of a certificate of practice under the Architects Act and, if the Council does not direct the issuance of the licence or the certificate of authorization, the Council shall give its reasons therefore in writing to the Joint Practice Board and to the applicant for the licence or the certificate of authorization.

Referral of dispute to Joint Practice Board

(4) Where a dispute arises between an architect and a professional engineer or a holder of a certificate of authorization as to jurisdiction in respect of professional services, the Registrar may refer the matter to the Joint Practice Board and the Joint Practice Board shall consider the matter and assist the architect and the professional engineer or the holder of the certificate of authorization to resolve the dispute in accordance with the rules in section 12.

Commencement of proceedings

(5) Proceedings shall not be commenced under this Act in respect of a matter

mentioned in subsection (4) except upon the certificate of the chair of the Joint Practice Board that the Board has considered the matter and has been unable to resolve the dispute.

Certificate

(6) The certificate of the chair is admissible in evidence in all courts as proof, in the absence of evidence to the contrary, of the facts stated therein without proof of the appointment or signature of the chair. R.S.O. 1990, c. P.28, s. 47.

Annual report

48. (1) The Council shall make a report annually to the Minister containing such information as the Minister requires.

Idem

(2) The Minister shall submit the report to the Lieutenant Governor in Council and shall then lay the report before the Assembly if it is in session or, if not, at the next session. R.S.O. 1990, c. P.28, s. 48.

Application of Corporations Act

49.(1) The Corporations Act does not apply in respect of the Association except for the following sections of that Act which shall apply with necessary modifications in respect of the Association:

1. Section 81 (which relates to liability for wages).

2. Section 94 (which relates to auditors) and for the purpose, the Minister shall be deemed to be the Minister referred to in the section.

3. Subsection 95(1) (which relates to the auditor's qualifications) and, for the

purpose, the subsection shall be deemed not to include,

i. the exception as provided in subsection 95(2), and

ii. the reference to an affiliated company

4. Section 96 (which relates to the auditor's functions).

5. Subsection 97(1), exclusive of clause 97(1 (b), (which relates to the auditor's report) and for the purpose, the Association shall be deemed to be a private company.

6. Subsection 97(3) (which relates to the auditor's report).

7. Section 122 (which relates to the liability of members).

8. Section 276 (which relates to the holding of land) and, for the purpose, the Minister shall be deemed to be the Minister referred to in the section.

9. Section 280 (which relates to making contracts).
10. Section 281 (which relates to power of attorney).

11. Section 282 (which relates to authentication of documents) except in respect of information from the records required to be kept by the Registrar

12. Section 292 (which relates to validity of acts of directors).

13. Section 297 (which relates to directions by a court as to holding a meeting).

14. Section 299 (which relates to minutes of meetings).

15. Section 302 (which relates to books of account).

16. Section 303 (which relates to untrue entries) and, for the purpose, the section

shall be deemed not to refer to section 41 of that Act.

17. Section 304 (which relates to the place of keeping and the inspection of records) and, for the purpose,

i. the section shall be deemed not to refer to sections 41 and 43 of that Act, and

ii. the Minister shall be deemed to be the Minister referred to in the section.

18. Section 305 (which relates to inspection of records) and, for the purpose, the section shall be deemed not to refer to section 41 of that Act.

19. Section 310 (which relates to investigation and audits).

20. Section 323 (which relates to evidence of by-laws and certificates of amounts due).

21. Section 329 (which relates to appeals).

22. Section 331 (which relates to untrue statements) and, for the purpose,

i. the section shall be deemed not to .refer to regulations made under that Act, and

ii. the Minister and the Deputy Minister to the Minister shall be deemed to be the Minister and the Deputy Minister referred to in the section.

23. Section 332 (which relates to orders by the court) and, for the purpose, the section shall be deemed not to refer to creditors.

Interpretation

(2) For the purposes of subsection (1), a member of the Association shall be deemed to be a shareholder. R.S.O. 1990, c. P. 28,s.49.

REFERENCES

[1] Gordon C. Andrews, "Canadian Professional Engineering and Geoscience: Practice and Ethics"

[2] D.L. Marston, " Law for Professional Engineers".